J

I KEPT NO DIARY

I KEPT NO DIARY

60 Years with Marine Diesels, Automobile and Aero Engines

by
Air Commodore F. R. (Rod) Banks
CB., OBE., Hon. FRAeS., FIMechE.,
C Eng.

Airlife Publications

England

Airlife Publications

7 St. John's Hill, Shrewsbury, England.

Printed by Livesey Limited, Shrewsbury, England.

To Con, who has given me more than fifty years of happy marriage and looked well after our family.

The Author at the age of 70. Hawker Siddeley Photograph.

Preface

This autobiography is not strictly a technical treatise, though certain aspects of my work might be of historical interest to those who have been similarly involved. I have for more than sixty years been engaged in the development and running problems of the internal combustion piston engine (marine diesels, automobile and aero engines). Then, in 1944–46, I was responsible for the future planning of the aero gas turbine engine.

Anyone so closely involved in his professional life with two major developments of universal importance can consider himself exceptionally lucky. I was associated with the introduction and use of the anti-knock compound, tetraethyl lead, and then, some eighteen years later, the aero gas turbine. This is the luck for which, when it comes, the recipient can give thanks to the Almighty because any job, anywhere, requires an element of luck to complement what professional skill or ability he may have.

The introduction of tetraethyl lead in the States, where it was discovered as a very potent anti-knock compound, was attended by a considerable resistance movement from some large and pre-judiced oil companies and also from Government oil cartels elsewhere. In Europe, excess alcohol (from the vineyards) and benzole (from coal and coking ovens and the steel plants) were Government monopolies, and these components were added to gasoline (petrol) to improve its anti-knock value. This alcohol/benzole monopoly was not entirely satisfactory because the quantities available varied according to, say, the grape harvest and/or steel production. Therefore, the knock rating (anti-knock value) was often inconsistent. But, with the lower compression ratios and lower specific power output of automobile engines between the two World Wars, and few motor roads in Europe, these differences were not always so apparent and caused little damage except possibly to some sports car engines. There was also the question of the health hazard of tetraethyl lead which had to be contested and argued in a number of countries. However, from its start and because of the meticulous and rigid controls and attention to the health of the individual workers in the lead manufacturing plants and others handling it, tetraethyl lead has had a clean bill of health for half a century. It has permitted the design of more efficient (high compression) engines to take advantage of gasoline (petrol) of higher and consistently maintained anti-knock value; not only in producing fuel of a standard quality but raising that standard from time to time, to meet engine demands, at a relatively moderate increase in refining costs.

In a different way, but with a profound influence upon aviation, came the development of the gas turbine. In about three decades it has virtually eliminated the piston engine, except for the small private aircraft. The aircraft designer, and the aerodynamicist at his right hand, have been given a prime mover of practically unlimited power and thrust, whereas the piston engine was relatively restricted in power which could only be advanced by somewhat small increments after considerable development effort. The aeronautical engineer's 'field day', to take advantage of the thrust potentialities of the gas turbine, resulted from the advances first disclosed when the allies went into Germany and found important developments in the aerodynamics field, like the swept wing. Since then have come further improvements in structures and materials of construction – such as titanium.

These aerodynamic advances could only have been exploited by the invention of the gas turbine – the turbojet, the propeller turbine and the fanjet; particularly the turbojet, which has taken the aircraft up to the high subsonic (mach 0.9), the supersonic (mach 2.2–2.5) and to the edge of the hypersonic (mach 3). From the original Whittle engine to those that have followed it over the last thirty years, the specific fuel consumption (sfc), in pounds of fuel per pound of thrust per hour, has been reduced to about a third (from approximately 1.5 lb/lb/t/hr to 0.5 lb/lb/t/hr) by a combination of high pressure ratio, improved component efficiency and higher turbine working temperature. More recently, the ducted fan has still further reduced the specific fuel consumption and noise levels in some engines by increased bypass ratio, still higher pressure ratio and turbine temperature to give an sfc of near to 0.45.

The Concorde, which has four straight, two-spool, turbojets of the axial type, flying at Mach 2 at 50,000 ft, requires about 145,000 equivalent thrust horsepower. Under these conditions, it has an sfc of about 0.337 lb/ethp and a pressure ratio of 88:1.

Aerospace, including rockets, thus covers such a wide spectrum in the advanced technologies, electronics included, that any country capable of exploiting them can expect to be with the world leaders.

I started to learn the engineering of my day when an apprentice of only fourteen years of age. My father was an engineer and worked with the early automobile and aviation engines. Helped by his knowledge, I often went with him to automobile races and speed trials and to some of the earliest aviation meetings.

Though not educated in the formal sense, my engineering apprenticeship and the continuous learning process which goes with all advances in engineering has stood me in good stead by its complete involvement.

This is not an adventure story, since, in any case, I could not write sufficiently dramatically or well. It might, however, be described as a personal record of events in which an engineering life has involved me, a life that could be considered a vocation rather than just a job.

I chose the title *I Kept No Diary* because, while on active service at sea in the First World War, we were enjoined not to keep diaries in the event of capture. In any case I am not of the diary-keeping fraternity.

F. R. B.,
London, 1978.

Acknowledgments

Even with the encouragement I have had from friends, I doubt if I would have completed these memoirs without the help of Mrs Louie Bond. She has typed and retyped these pages many times, intelligently reading my 'shorthand' and making helpful suggestions as to editing. I owe her a great debt and give many thanks.

To Alan Bramson and Group Captain E. 'Teddy' Haslam (Head of the Air Historical Branch [R.A.F.] M.O.D.) I can only say an inadequate 'thank you' for their personal help and considerable expenditure of time in advising on and improving the book.

Others who have helped through conversations, suggestions and information are: John Allen; T. R. A. (Ray) Bevan; Jim Corfield; S. D. Davies; Maurice Elliott; Errol Gay; Michel Hedde; Robert (Bob) Hotz; R. W. Harker; The Associated Octel Co. Ltd.; R. V. (Bob) Kerley; Philip Lucas; Dr Eric Moult; A. R. Ogston; J. J. Parkes; M. Ramsden, Sam Savoca of TRW; Derek Wood. Finally, Rolls-Royce Ltd. and Hawker Siddeley Aviation Ltd., and all my friends there.

To Arnold Nayler, Mrs Eva Dane and Michael Fitzgerald of the Library of the Royal Aeronautical Society, I give thanks for their unstinted personal help in searches to confirm references and dates, etc.

F. R. B.

Contents

1 *Early Days*

I was born the eldest of six, three brothers and three sisters, at Kings Heath, then a village outside Birmingham and now I believe an integral part of that City. It was by chance that I was not born a cockney, since we lived in North London, but my father had to go to Birmingham, temporarily, to wind up a failed engineering firm there.

My father was a capable mechanical engineer and a good mechanic. In other words, he could make what he designed, which included anything from a rotary printing press to a machine that automatically produced air-gun slugs.

His family originally came from Dorset and the ancestry can be traced back to Lady Banks, the defender of Corfe Castle against the Roundheads. Two cousins of my father were Sir William Barrett, one time professor of mathematics at Dublin University, and A. A. Milne, the author. Sir William and Sir Oliver Lodge used to correspond a great deal on the possibility of life after death. Both were believers, but apparently had little or no manifestation of it. Another ancestor was Sir Joseph Banks, who sailed with Captain Cook to Australia and, with him, founded Botany Bay.

My mother came from Essex, where her father was a country squire. The family, of thirteen children, lived in East Mersea Hall on Mersea Island, which faces Brightlingsea on the River Colne. I never knew my maternal grandfather since he died at a relatively early age and before my mother married. He was William Cockrell, whose ancestry was Norman. His death left my grandmother to fend for herself with the thirteen children, which she did, helped by good friends and, later, by the efforts of the boy members of the family who found jobs for themselves in the City. One, Leonard, went to sea as an apprentice in sail, eventually to become a ship's captain, gaining a D.S.O. in the 1914–1918 War when in charge of an armed yacht. Later, at the age of 70 years, in the Second World War, he commanded convoys. Other uncles went into banking.

My father's father was a bank manager, and my paternal grand-mother who, incidentally, was thirteen years older than my grand-father, was a Cowdell Barrett and related to the Barrett family (of Wimpole Street). While my mother's father was alive he was responsible for the living of the church next to the Hall, and the incumbent for a time was the Reverend Baring Gould who also had a large family. Baring Gould wrote the two hymns, 'Onward, Christian Soldiers' and 'Now the Day is Over'. He also wrote the first real novel of Essex, 'Mehalah'.

As a boy, my father used to go to Norfolk for holidays and on these occasions 'apprenticed' himself as a helper to the local blacksmith; assisting on one occasion to instal a fireplace in the house at Burnham Thorpe where Admiral Lord Nelson was born. He became a good blacksmith and in later years used to make fire-irons for his friends, and on one occasion an ornamental iron gate for a friend's Norfolk manor house.

After the Birmingham interlude, we returned to London and lived in the northern suburbs, first at New Southgate and then at Crouch End, which saw the start and finish of my schooling. I went first to a small preparatory school and then to a good grammar school. I was not particularly fond of school but tolerated it, passing through the various classes and grades without any distinction – always about the middle. I was little good at games but a fair boxer and could take on, and sometimes beat, older boys in fights. The only prize I won at school was that given by the headmaster when the cadet corps, of which I was a member, went to the schools' summer meeting at Bisley. I was in the school shooting eight. He gave me this small cup for the best aggregate score for an individual boy's shooting in all the events. We were quite highly placed, eighth, I think, in the Ashburton Shield competition in 1912. At Bisley, we were inspected and given a talk by Field Marshal Lord Roberts.

In the late nineteenth century, my father and grandfather cycled a great deal all over England and some of it on penny-farthings. But my father early became interested in the motor-car and drove various pioneer makes such as a Cannstatt-Daimler and De Dion Bouton. He went to the States in 1898, the year of my birth, and bought some Whitney steam cars on behalf of his Company, Brown Brothers, then cycle manufacturers and factors. Over many years they have been the leading motor accessory factors. My father drove one of these Brown-Whitney steam cars in the first Thousand Miles Trial in 1900, sponsored and organized by the Automobile Club (later, Royal), and received a non-stop certificate. He did in fact stop a number of times to repair the boiler and fill up from streams and ditches, sometimes obtaining water from reluctant and suspicious householders, but making the control points on time.

At the time of the Boer War there was a general urge to know about the handling of weapons, and volunteers were training in anticipation of joining the Army. Stemming from this was an interest in target shooting with air-guns, and my father was a keen experimenter with this particular weapon. He possessed some powerful air rifles, British and German, quite lethal in fact at twenty or so yards, and devised a more efficient slug, the Bulfin. He designed and made the machinery to produce it in the small machine shop he owned at Dalston in the East End of London.

I was intensely interested in aeroplanes and, at the age of fourteen, designed and built a large rubber driven model of an inherently stable monoplane, copied from the German Etrich *Taube*, which had competed in the Circuit of Britain Air Race in 1911. This later won me a bronze medal in the model section of the aero exhibition of 1912 at Olympia.

Shortly afterwards I left school and went as an apprentice to my father's machine shop in Dalston, learning to handle the normal tools of the period, the screw cutting lathe and milling and drilling machines. He overhauled automobiles and their engines and also made bits and pieces for some of the flying people at Hendon, where I often went with him. Here I made my first acquaintance with the Gnôme rotary engine. He also supplied special fittings to S. F. Cody from time to time and in fact gave him a good deal of material help. Cody, American born, became a naturalized British citizen and the first man to fly in Great Britain, in 1908, piloting a machine of his own design. I assisted in the overhauling of engines, and in my spare time tuned the motorcycles of my more affluent local friends. Thus I learned something about the internal combustion engine. When I was about eight years old my father took me to the first speed contest at Brooklands Track, between S. F. Edge on a Napier and F. Nazarro on a Fiat, enormous engines with great pots of cylinders.

It was in 1912 that an Australian, a Mr Craig, bought two or three Brown cars from my father. Craig owned a large store in Melbourne and was also a racehorse owner. My father took him to Brooklands one day and he became very enthusiastic about motor racing, but not as a driver. He was a massive man and well over six feet tall. On my father's advice he bought a 27.9 hp Benz, the successful racing car of the day, and took on a driver, Wilkinson, to compete at Brooklands. The finish of Craig's interest in motor racing came when Wilkinson took the banking before the Railway Straight at the wrong angle, went over the top and hit a telegraph pole. He was unconscious for nearly a month.

I drove my first motor-car at the age of eight in a country lane near Mersea Island, my father instructing, and I seem to have been driving cars ever since. There were no synchromesh gears in those days; cars had crash gears and leather-faced cone clutches, though the multi-plate type was coming in. One often had to jack out the cone clutch with a wedge of wood, when the car was not in use, to prevent it sticking and also to treat the clutch leather with Collan-oil. Engine oil was used by Rolls-Royce for the same purpose at the time.

Through my father's friendship with the Farman brothers, I had my first flight in a Henry Farman at Hendon in 1913, piloted by Chevillard the well-known Farman test and stunt pilot. In the same year, I saw Pégoud loop-the-loop at Brooklands in his specially rigged and strengthened monoplane. My father got Wilbur Wright his first mechanic when he came to France to demonstrate his machine at Le Mans and Pau. This man, George Mills, the son of an Australian gold millionaire, did not last out Wright's French visit, since Wilbur insisted that he sleep on straw alongside the machine in the shed where it was housed and George thought this both unnecessary and most uncomfortable. So, my father had to get Wilbur a tough and competent French mechanic.

My father was below average height, but was broad built and physically very strong. I once saw him lift, quite easily, the near side rear of an early car on to a jack, in order to change a tyre. This must have meant lifting a deadweight of some hundred or more pounds. He was, unlike so many of small stature, a quiet man and never boasted or blustered. He never raised a hand to us when we were children and left the running of the house and family to my mother, though he was master and she deferred to him at all times. She was good at housekeeping and in those days, even on my father's relatively modest earnings, we could afford plenty of help in the form of a cook and a housemaid. For some years, a German lady lived with us and taught us to speak German and write it in the Gothic characters. Now all forgotten, alas!

I would not say that my parents were great family people. My mother and father were very close to each other, and though they looked after us well it could not be said that we necessarily came first and foremost in their thoughts, except to do the right thing by us. Though my father was master of the family, my mother got him to do things that he would not, I suspect, have done if left to his own devices. They made several changes of home, not having lived very long in the previous ones, purely under my mother's pressure. She was restless in this respect.

After a year at Dalston, I went to Lowestoft and continued my apprenticeship at an ice-making and refrigeration company – The East Anglian Ice & Cold Storage Company. This Company had been formed at the end of the last century by my mother's eldest brother, Fred Cockrell and my father with the Australian gold millionaire, Mr Mills (father of George Mills, Wilbur Wright's first mechanic at Le Mans) and a Mr Rice. While there, I took tuition in engine design and drawing office practices from Stanley Rose, the chief designer of J. W. Brooke & Co. Ltd., who designed engines for their high quality launches.

At Lowestoft, I learned about refrigeration on one hand and helped to 'mechanic' Brooke racing motorboats for various enthusiastic and wealthy owners. Hollingsworth, of Bourne and Hollingsworth, had some boats built for him by Brooke; Cordon Rouge was one (named after his favourite champagne), a twin screw, hard chine, vee bottom, design. Another boat was Crusader, a Thornycroft type, single step, hydroplane. The interesting feature of these boats was the engine, designed by Rose, an eight cylinder, 90° vee job of 300 bhp. The boats had speeds of between 35 and 40 knots, which was quite fast in those days. Brooke's chief mechanic, Bob Luen, whose father was employed at the Ice Company, was one of the most capable and knowledgeable engine mechanics I ever met. Not only did he know his engines and could tune them effectively but he was a real master of all tools, machine and hand. He could make almost anything and, like my father, undertake blacksmithing when needed.

I had most interesting and, to me, exciting interludes while at Lowestoft. The first was a round trip from Cardiff to Philadelphia and New York as a greaser in the engineroom of a cargo boat. The second, and more exciting, was to go as an improver mechanic with Bob Luen, and others, to help look after Brooke's racing boats whose owners were competing at the International Motor Boat meeting at Monte Carlo. This was in the Spring of 1914.

In the pre-war years, such International meetings were generally held at the end of the Season on the French Riviera and practically marked the closing down of Monte Carlo until the following Autumn – the hoteliers moving to Biarritz and Deauville and other resorts for the summer months. I don't know when these motorboat meetings were first instituted, but this one, which was to prove the last before the 1914–18 War, was a real occasion for me. We were put up at the Bristol Hotel and were also made honorary members of the Sporting Club. I was therefore very lucky to have a glimpse of Monte Carlo, and incidentally Nice, Cannes and Paris at the height of their rather exclusive existence.

There was a boat built by the French firm of Despujols which was known as *Jumping Jesus*, since it planed right back on its stern with its bow high in the air and, in a chop, leaped from wave to wave. It was at this meeting that I saw a large bearded mechanic, who was with one of the French entrants, take a stripped timing wheel, mark out the teeth on a steel disc or blank he had got from Nice and then, filing throughout the night, he formed all the teeth for a new camshaft gear which he fitted to his engine in time for a race the next day. In those days, a number of marine engines had exposed timing gears, among them the Brooke engines which had the half speed (camshaft) wheel of compressed and impregnated fabric sandwiched between bronze side pieces. They were quite silent and reliable in operation.

During my time at Lowestoft I went on an Icelandic fishing trip from Grimsby. This was a very tough but most interesting experience. The living conditions on board were crude to say the least and quite unlike those in modern trawlers. These earlier vessels had none but the most rudimentary accommodation and conveniences. Since one had largely to live in one's clothes, sleeping in long woollies off watch, there was the chance of becoming verminous after three or four weeks at sea. In fact, being bitten by fleas and bed bugs on that particular trip completed my discomfort. On top of this, bad weather decided me that deep sea fishermen earned their pay.

2   The 1914–1918 War

The start of the 1914–1918 War found me at Lowestoft and within the first week I had tried, successively and unsuccessfully, to join the Royal Flying Corps and the Royal Naval Air Service. Both were suspicious of my age and requested a birth certificate which, because of my sixteen years, hardly qualified me for entry into either Service as a pilot. Then I heard of a certain Army Major who was recruiting motorcycle despatch riders for the Expeditionary Force, and particularly wanted those who spoke German. I was then fluent in that language and rushed to London to be interviewed, but was again turned down on account of my age – probably a very lucky escape.

In desperation, since everyone seemed to think the War would be over in weeks, I joined the Royal Navy at Lowestoft as a T124 man, meaning someone not necessarily qualified or experienced in matters to do with the sea or the Navy in particular. Forewarned by my previous attempts I gave a false age of eighteen but in the event the Senior Service did not query it.

Too young to realize it at the time, in that summer of 1914, and having no idea of the events which were to follow the assassination at Sarajevo, I obviously did not appreciate that my relatively carefree existence was gone. Life was never to be quite the same for those of us who were lucky enough to survive. But, almost without realizing it, one slipped easily and effortlessly into Service life and routine; mainly, I think, because there was more discipline in the home and with my cadet corps background. True, there was the Service discipline and toughness not seen today, but one took it in one's stride. And since there had been no previous experience of large scale war, everyone wanted to see active service.

Looking back at this period of my life, when many of my age were still at school, and after being twice blown up by mines and also helping to dismantle one, such incidents only appear exciting in retrospect. The real excitement was of being in the War.

When my parents heard that I had joined the Service, and they only learned of it when it was a *fair accompli*, my father, characteristically, said nothing but appeared pleased. But my mother was naturally worried and wondered whether I could not have served the country as well in the Boy Scouts! But I was already an 'experienced and travelled' improver mechanic and had had my first alcoholic drink, a Bass No. 1 Barley Wine, pretty strong for anyone, so I wasn't very receptive of the Boy Scout idea.

The Lowestoft naval base was on the South Pier and mainly concerned with minesweeping and patrolling operations; with deep sea trawlers, coming from Grimsby, Immingham and elsewhere, to be fitted out as minesweepers in the local yards. As early as 1907 the Admiralty had, at the suggestion of Admiral Lord Charles Beresford, then C-in-C of the Channel Fleet, organized a special Trawler Section of the Royal Naval Reserve. This was, I think, actually put into effect in 1911.

The S.N.O. and Commander of the Base was a Post Captain, Alfred Ellison, a strict disciplinarian. He needed to be since the fishermen trawler crews of those days were a pretty hardbitten lot. With the public houses open at all hours there were always a dozen or so trawlermen at Captain's defaulters every morning in the early weeks of the War – some so violent that they often tried to go for the Captain and had to be handcuffed. A shore party in charge of a Lieutenant, R.N., went round the pubs. every night and arrested the obstreperous ones. But, tough as they were, Ellison was tougher and eventually got some semblance of order and discipline into the minesweeping flotillas, helped by the senior R.N. and R.N.R. Lieutenants commanding them.

The navigating officer of the Base was Commander Bruce, R.N.R., who was one of the Ship's Party of Scott's ship the *Terra Nova* on that ill-fated Antarctic Expedition of 1912. Wilfred Bruce's sister had married Scott and she used to visit the Commander at the Base, accompanied by a very young Peter Scott whom she seemed to be toughening up for his future life since he wore the minimum of clothes all through the year – and the winter months were very chilly at that most easterly part of England.

Bob Luen, myself and Jack Eastty, another Ice Company colleague, had all joined up at Lowestoft and were given the new status of Motorboatmen or Motorboat Mechanics, equivalent to Petty Officers, First Class. We were put in charge of some fast (Brooke) motorboats for running errands to vessels at sea and taking officers and others on boarding parties and such duties. Some of these motor launches were actual racing boats, and for many weeks didn't have a compass between them! We, literally, had to feel or smell our way out of harbour and around the sandbanks; thank heaven for their shallow draught! The three of us were responsible for their handling, upkeep, cleanliness and mechanical condition.

After a month or two of this I transferred to minesweepers and had a somewhat tougher time, finding myself in the water on two occasions owing to mine explosions. The first was when we were attempting to lever the sweep wire, which had overridden the 'gallus' block (pulley) and jammed. The sea was choppy, it was the early winter of 1915 and very cold. The sweep wire was slacked off by our sister trawler but there were two mines, broken from their moorings, in the sweep. Since we still had some forward way the mines came together in the trailing loop of the sweep wire and detonated. The resulting explosion blew me into the water. I had on oilskins and seaboots but, having swallowed a lot of water and the sea being very cold, with some concussion due to the explosion I was nearly gone. It was some minutes before a boat could be lowered to pick me up. Those who have experienced near drowning, with a lot of water in the lungs, will know the painful process of being brought round again. The second occasion was while we were trying to recover a mine from a fishing boat's net. It went off and blew three of us into the water, even though we were quite some distance away and had just cut the net adrift.

In the early part of the War, the British moored, bar type, mine proved most ineffective. Whenever one broke loose and floated free (if it did not sink as prescribed by the Hague Convention) it could be handled almost with impunity, since the tilting bar mechanism was usually corroded solid. On account of this unsatisfactory state of affairs, and with much evidence of the effectiveness of the German mine, a sweeper, the *Columbia*, was instructed to recover one and bring it back to Lowestoft for examination. I assisted a Leading Torpedoman from the *Vernon* to dismantle it, under the general direction of Lieut. Sir James Domville, Bt., R.N. One of the lead horns of this mine had been hit and dented at some time and, after removing it, having first cut the wires from the distributor to the detonator, we broke the sealed glass electrolyte or acid container in getting it out of the horn!

Domville, a somewhat wild man in his early days in the Service, was, I believe, asked to resign before the War owing to recurrent overstaying of leave, but their Lordships accepted him back at the beginning of hostilities. He had plenty of guts and shortly after the mine dismantling episode, when in command of a group of three or four trawlers, he engaged two German torpedo boats near the North Hinder off the Dutch coast. He met death by his own hand in a Service Club in London after the War – characteristically leaving a note for the Club secretary, apologising.

I was, later, in what must have been the earliest of the 'Q Boats', decoy vessels with concealed armament to tackle U-boats. In this particular activity, ordinary fishing smacks (sail with auxiliary engines) were used to meet German attacks on the North Sea fishing fleet. They were using the smaller and older type of U-boat, which would surface in the middle of the fishing fleet off, say, Smith's Knoll, NE of Yarmouth, then a demolition party would board one or two smacks and place small bombs to blow out the stem and stern posts, torpedoes being too valuable for this purpose. Our particular smack, the *Kingfisher*, had a 12-pounder gun concealed in a mock companionway hatch which could be collapsed very quickly, the gun being already loaded and manned. The drill was to hold fire until the submarine was alongside, or practically at point-blank range. It was of little use otherwise, since the U-boat might get away and, being forewarned, report this unorthodox method of warfare, which could lead to reprisals. Once it was decided to engage, down went the hatch covering and the gun was brought to bear on its target, at the same time the White Ensign was run up.

We never had the luck to tackle a submarine while I was on board but underwent the rather frustrating experience of watching a submarine's demolition crew boarding two smacks without being able to do anything to help. The smacks were between us and the enemy who, in any case, was too far away for us to be sure of a disabling hit.

One day, in January 1916, I was summoned to the S.N.O., Captain Ellison, who said that he had decided to recommend me for a commission as a Sub-Lieutenant in the Royal Naval Volunteer Reserve (R.N.V.R.). This was quite unexpected but when I thanked him he grunted and said all he wanted to hear was that nothing in my subsequent career would give him cause to regret his decision or suggest that his judgement had been questionable. My commission came through within a week and I was commanded to report on board HMS *Hermione* at Southampton. The 'Hermy One', as she was called, was the R.N.V.R. base ship and here we were taught the duties and behaviour expected of officers and gentlemen.

I had been on board less than a month when a notice was posted asking for applicants to serve in the Eastern Mediterranean, and particularly those with some experience of motor launch handling. I was one of three chosen. We were given a week's leave to get fitted out with tropical kit, which included the heavyweight solar topee that was then the official headgear for the tropics or near tropics. This was in fact my first leave in a year and a half of war service.

Our sailing orders instructed us to join a cargo vessel, the *Inkonka*, at Avonmouth, loading ammunition and explosives, with a deck cargo of aero-engine lubricating oil, all destined for Port Said. The Master, a man of about fifty years, had a sense of humour, though he was somewhat lugubrious. The sinkings of shipping in the Mediterranean by German submarines were then high and the Captain kept speculating at mealtimes where the safest place (sic) would be for a torpedo to hit, generally deciding that it would be the engineroom, much to the irritation of the Chief Engineer. It so happened that an hour before our sailing time, a fire broke out in that very place somewhere about the forward bulkhead. There was considerable smoke, making it difficult to fix the exact location; and the Captain and the First Officer were worried about the effect of an overheated bulkhead on the ammunition and explosives in the main hold on the other side. So was the Harbour Master who had a large number of other ships under his care. He ordered a tug to get us out quickly and told us to anchor as far as possible from Avonmouth and the busy part of the Bristol Channel. The Captain kept his head, murmuring that he was not a particularly religious man but never thought he might emulate Elijah and go to Heaven in a Chariot of Fire.

Eventually, after two hours, the fire was brought under control; an overheated bearing in a ventilating fan was the cause of trouble, firing the lubricant which in turn set fire to some wood shuttering.

I didn't after all command a motor launch but was appointed First Lieutenant of an armed sand hopper in the Suez Canal, based at Ismailia. The Commanding Officer was a Royal Indian Marine Lieutenant-Commander, Marsden. Our job was to patrol the Canal, so as to be available at points where the Turks might cause mischief. The Turks had in fact previously sunk a cargo ship in the Little Bitter Lake by placing mines in the Canal. They humped two of these on camels from the railhead to the Canal and the ship had hit one. Luckily, it had settled fore and aft in the channel and did not obstruct it seriously, nor was it more than about half submerged, though some days of strenuous effort were needed to get it refloated.

The other mine was located and proved to be one of a consignment of French mines sold to the Turks before the War. In attempting to recover this mine a line was secured to it and the other end fastened to a donkey whose Arab owner was instructed to go very carefully and pull it slowly into shallow water. Unfortunately, the line was too short and the mine exploded as soon as it grounded and tilted over, killing both man and beast.

The Australian troops of the Anzac Force, a pretty tough bunch, had been brought to the Canal Zone after the evacuation of Gallipoli, to help defend the Canal against any possible Turkish attack. There was an Australian Field Gun Battery near Kantara which used to take delight in firing salvos whenever we passed on patrol; the shells, at relatively flat trajectory, once removed our wireless aerials. Again, their sentries on the Canal bank would often challenge us at night, firing at the same time in order to hear the bullets 'ping' off the side of the ship.

All troops were given leave at intervals to Port Said and Alexandria in order to cool off from the hot desert air of the Canal Zone. Arabtown was put out of bounds and, if I remember rightly, there was a 'no drink' order until about midday and then only fairly weak beer was available. But the Australians seemed to get plenty of drink in them by about breakfast time and the Assistant Provost Marshal had his work cut out to keep order. In fact, he was always escorted by six or more of the toughest and heaviest military police he could muster to do his job at all. It was my impression that a large proportion of the Australians of that period came from the Outback as distinct from the majority in the Second World War who probably originated from the large communities in the cities. But in both cases, the men were tough and fearless in battle, though in the first World War they were less disciplined. The sacking of the Waza, the red light district of Cairo, where they threw prostitutes and pianos out of the windows, is now part of the history of the Australian 'occupation' of Egypt.

I was seconded from the armed hopper on two or three occasions for spells of duty in each one of the two Royal Indian Marine ships, *Dufferin* and *Hardinge*, and twice met Lawrence, whom we took to Suez from the desert on his way to Cairo and back. This extraordinary but quite unfathomable man was much less impressive in appearance than I had imagined him to be, not realizing until then that he was below average height. There was no doubt, however, of the force of his personality, though he could on occasion be somewhat rude and overbearing.

It was in the *Dufferin* and *Hardinge* that I first encountered, in a big way, the Senior Service's favourite drink, pink gin and, after sampling too many, the prairie oyster. I never could compete with wardroom drinking in the mid-morning in the Red Sea at a shade temperature of around 100°F. Sometimes, when on patrol, to intercept gun-running dhows from the African Coast to Jeddah, it was necessary to turn for a spell and steam against the wind in order to give some relief to the ship's company.

It was on these tours of duty that I first saw Karachi and Bombay and also visited Colombo (Ceylon) and Calcutta. Karachi is very different now from what it was then, having grown from a small fishing port to a large town. Though I like India and the Indian people, I have never cared for Calcutta. It is, I suppose, because of the awful poverty which is in stark contrast to the wealthy commercial parts of the city.

During my time on Canal defence we went into the desert on various missions and on more than one occasion were charged with taking sacks of gold sovereigns to Lawrence's agents for payment to the Arab leaders for their services against the Turks. On two occasions, I was among those feasted by a Sheikh in typical fashion; a whole sheep on a dish with rice and everyone grabbing pieces of greasy meat, in a stifling atmosphere and a temperature of 120°F or more. Luckily, I was too junior to be offered an eye – my Commander had that honour!

We went to Port Said at frequent intervals in our armed sand hopper, where I struck up a friendship with Dacre, C.O. of the R.N.A.S. Seaplane Base, and Banks-Price, Brooke and other pilots and personnel there. I also met, or rather encountered, the famous Commander Samson of the aircraft carrier, *Ben-My-Chree*. I was for a very short time on patrol up the coast in the *Ben-My-Chree*, which was a former Liverpool/Isle of Man passenger steamer converted into an aircraft carrier for seaplanes. Commander Samson was below average height but a tough, fiery officer and a pilot of considerable ability and distinction. We used to cruise off the Eastern Mediterranean coast as far north as the Gulf of Alexandretta (now Iskanderon). The *Ben* was eventually sunk by Turkish heavy battery fire from the mainland as she lay at anchor in the island harbour of Kastelerizo. I had left for home before this happened.

Samson used to cause much amusement among the ship's companies of other naval vessels when he entered Port Said, with the crew dressing ship and a motley crowd, composed of anyone who could play an instrument, playing the ship in to her anchorage – on drums, tin whistles, mouth-organs, and even a violin.

Since there came a fall off in Canal patrol duty I requested and was given a temporary appointment to the Seaplane Base, assisting in engineering duties and also going as observer on flight patrols up the coast from Port Said in Short Seaplanes. This was usually a very early morning job, since the Short with a maximum load of fuel and a few lightweight bombs, machine guns and ammunition, could hardly take off in the full heat of the day.

We were ordered to patrol off El Arish, some ninety miles from Port Said, where the Turks had an army headquarters, to spot for a monitor firing on the town. I think this class of monitor had 9.2-inch guns. Anyway, while we 'dotted and dashed' the overs and unders on our wireless, with no registered hits in an hour of firing, most of the shells ricochetting far into the desert, we were attacked by a German Rumpler aircraft. We received several hits on the fuselage and wings, but nothing serious until it took us almost head on, from above on the starboard bow, and put a burst into the radiator which on the Short stood high and ahead of the centre section, luckily missing the pilot and myself.

The Sunbeam engine eventually seized up and the pilot put us down rather heavily near the monitor, pulling a float off. We hung on until a boat came to pick us up and put us on board, where we were given breakfast while our clothes were drying. Sitting next to a Lieutenant R.N., I rather innocently asked why no notice was taken of our signals, since nearly every shot had bounced miles into the desert. He looked at me hard and then said, 'You reserve officers know nothing about gunnery'. He was in fact the gunnery officer and, obviously, I had lacked tact. But I too had done a gunnery course at Whale Island! Admiral Sir Rosslyn Wemyss, then C-in-C East Indies Station, was however always most charming and understanding and particularly decent to us reserve incomers.

3 Return to Home Waters

Shortly after this episode Dacre tried, unsuccessfully, to get me into the R.N.A.S., but I was eventually ordered home to be given command of a Motor Launch (ML). I commissioned her at Southampton and was then ordered to Immingham on the Humber. Anyone based at that fishing port in the winter, after some time in the Eastern Mediterranean, can realize the let-down I suffered at the hands of their Lordships.

MLs, which were used for anti-submarine work, later carried ten D-type depth charges with a 13-pounder short-barrelled gun forward. This gun was difficult to fire accurately, unless in a dead calm, since it needed three to handle it – the trainer and layer, one rating to traverse and one to load. The ML could pitch and roll equally well, so unless one was quick the shell could either hit the water near to the vessel or disappear over the horizon. The 13-pounder was eventually removed from a number of these vessels and replaced by the more handy shoulder trained 6 pounder. In some cases, with the Dover Patrol, Lewis guns were mounted for anti-aircraft purposes. There was also the lance bomb, consisting of an explosive head of about 17 lb of TNT attached to an ash handle or shaft. The idea was to launch it like a javelin, but it was so heavy and awkward to handle that it proved as lethal to the thrower as to the enemy and they were later withdrawn. We used them to dynamite fish, particularly when there had been no submarine activity to warrant dropping depth charges.

MLs were twin screw, wood built, vessels constructed in quantity by the Electric Boat Company of Bayonne, New Jersey, U.S.A. A number were assembled in Canada from Electric Boat components shipped there from the U.S. There were some 550 ordered by the Admiralty, the first fifty being 75 feet and the rest 80 feet in length. They were quite good for their purpose and handled well, even in a heavy sea. But it was rather like riding a roller coaster and one fared better if reasonably immune from seasickness. The engines were

Author's ML in the North Sea during World War I.

MLs in a rough North Sea.

interesting: the alleged power was 220 bhp (each) at 460 rpm and they were open crankcase jobs with drip feed lubrication via slinger rings, to plain bronze main bearings and big ends. They had automatic inlets and cam and pushrod-operated exhaust valves with a sliding camshaft for directly reversing the engine, initiated by compressed air. There was, therefore, direct coupling between engine and propeller shaft. Sometimes, due to hurried wartime workmanship, the keying of the exhaust cams to the shaft was not always well done and an order to reverse the engines was rendered ineffective owing to the cams sliding off their keys!

Great care had to be taken to avoid pieces of cotton waste getting caught in the adjustable dripfeeds to the big end slingers since, if one feed was diverted, this would quickly cause a complete seizure of the particular big end on its pin bringing the whole engine to a dead stop.

The MLs had a top speed of 19 knots and their range was about 1,000 miles at 15 knots. They carried 2,000 gallons of petrol (gasoline) and were, therefore, potentially lethal affairs because the fuel tanks, carried aft, were often strained by the weaving of the hull in bad weather. This resulted in leakage of neat petrol into the bilges. We lived in an aura of petrol fumes much of the time, so that when the Valor paraffin stoves (for cabin heating) were lit in harbour it needed little imagination to appreciate what sometimes happened! Generally, the resulting explosion blew off the cabin deck which, literally, acted as a safety valve; creating the minimum of damage if there was not too great an excess of petrol in the bilges to spread into an uncontrollable fire. I remember, in the Suez Canal Zone of all places, the surprise of the ML crews at Ismailia upon receiving barrels of sand *from home* to put out petrol fires!

After a spell at Immingham, I took command of another ML on the Tyne. It was a fine early summer day in 1917 when I arrived at Newcastle, immediately proceeding by electric train to Northumberland Dock where the ML flotilla lay. I was barely on board when the previous commanding officer, who was standing by to hand over to me, gave me a signal from H.M.S. *Bonaventure*, the Base ship, ordering us to proceed 'with all speed' to Whitby and locate and destroy three German mines which were reported drifting near the entrance of the harbour.

After reading the signal I was told of a thick fog from the Tyne to Flamborough Head, which had been there for some days and could remain for a week or more. The crew seemed rather surprized, therefore, when I acknowledged the signal and confirmed that we would proceed immediately. My predecessor, warning me of the hazards of the North East Coast, then departed for shore and his next command.

It only goes to show how ignorance can truly be bliss. I was granted a commission in 1916, when I was even too young to hold one at the time. I had enjoyed little experience of the sea except as a crew member in my uncle's yacht. Apart, therefore, from a short navigation course at the Royal Naval College, Greenwich, where some of our three weeks' concentrated work, which was supposed to be backed up by evening study, was interrupted by breaking bounds at night and going to the West End, I certainly knew very little of navigation.

We cleared the Tyne for Whitby.

In a fog, or at night with no navigational aids, in war, it was a rather tricky approach to the coast, let alone to make a narrow harbour entrance. If one misses the entrance to Whitby, there is the beach or the barely submerged Whitby Rock. But beginner's luck held, we made Whitby harbour before dark and moored safely alongside the quay.

At first daylight and the fog still with us, we somewhat gingerly left the harbour in search of the mines. After we had cruised around on one engine, two mines were, surprisingly, found within half a mile of the harbour. The difficulty then came when we secured them, by lines of different length to prevent them hitting each other, since the fog was too dense for us to stand sufficiently far off and sink or explode them by rifle fire without blowing ourselves up. So, we stood further out from the harbour entrance between the shore and the War Channel. The latter, a buoyed channel a few miles off shore, was kept swept and more or less clear of mines for all sea traffic and convoys plying up and down the Coast. We streamed the mines, like a sea anchor, and had to keep watch on them for a further two days and nights before the fog lifted. Then we released them and fired at a safe distance. Both were exploded, one detonating the other.

I was equally lucky in my navigation on another occasion. We had escorted a convoy from the Tyne to Flamborough Head in daylight but it was in such foggy conditions that I, as flotilla leader, received a signal to return to the Tyne forthwith. Leaving the convoy at dark and marking my estimated position on the chart we proceeded in line ahead up the War Channel, so much so that the third ML in line ahead hit a War Channel buoy, luckily a glancing blow, and partly stove in its bow. Running strictly on dead reckoning and to within seconds on my watch, when I judged myself to be abreast of the Tyne I turned to run in towards the entrance. Since it was pitch dark and still foggy, and having run my time and estimated distance, I stopped engines. There was a ground swell, indicating shallow water and proximity to the shore. Pointing the Aldis light at where I thought the coastguard station would be, I was most surprised and relieved to get an answering flash from almost above us. We were on the wrong, or Tynemouth, side of the harbour entrance so all we had to do was make a turn to starboard and, with the lead line, slowly complete the circle until a pier head came into view. Then, very gingerly, we crawled up the river to our moorings.

In those days we had hydrophones by which we could locate submarines but it was only possible to get the general direction with little or no idea of the range. A submerged submarine could be identified as distinct from a ship or, of course, a convoy, which swamped everything. But the Germans got wise to hydrophones, and as soon as we started our engines to close in, they stopped theirs. There was at that time a shortage of depth charges and we had to be fairly sure of our target before attacking. Later, in 1918, with no such shortage, destroyers were able to drop dozens, and even hundreds, to encircle a submarine.

Early one morning, some three or four miles off the Tyne and in another fog, we were in a rough circle of six MLs, hydrophoning. There had been considerable enemy submarine activity in the area. After a while, we heard the high speed beat of a number of propellers coming rapidly nearer apparently up the War Channel. Then, out of the fog and only a few cables length from us, came some destroyers in line ahead formation. They were not familiar, and as they disappeared into the fog we saw the German ensign!

Most of the ML crews had been fishing fleet men, but some were novices and even greener than I. The cook in my first command had been a bookmaker's clerk, who knew a lot about horse racing but nothing about cooking. Since the crew were almost mutinous about the food, and were threatening to do him bodily harm, I had to try and teach him the rudiments of preparing plain food. At first, he used to half peel the potatoes, leaving all the eyes in, and not fully cook them. They came to us with a 'Rockwell C hardness of 45'. Some of my better crew members were Scots fishermen, and they showed me how to cook herrings properly – split and fried in oatmeal, to reduce their oiliness. It was quite a sight to see these men get through a dish of herrings. Some would pick the fish up in their hands and then eat them off the spine, rather like playing a mouth-organ.

There was, in 1917, a call for volunteers for service in submarines. With the uncertainty as to whether or not the German High Seas Fleet would come out again, the submarine service was short of trained R.N. officers, those eligible being retained in the Grand Fleet. A number of us volunteered for this and were taken on, mainly for watch-keeping duties. It is significant that in the Second World War there were Reserve Officers in command of submarines, but as our period of training necessarily had to be short for a

temporary appointment we were obviously given minimum responsibilities. All I can say about this interlude is that the E-class
submarine was a most uncomfortable marine vehicle. The air got
so foul after a few hours that one lost all sense of taste and smell and,
except for appearance, could not distinguish between tinned fruit
and bully beef.

Our duties, from Harwich, mainly consisted of steaming into the
Heligoland Bight and sitting at periscope depth to keep watch on
enemy shipping movements. We would surface at nightfall, charge
batteries and wireless in what we had seen. The crew were allowed
on deck in twos and threes, so that in case of a sudden emergency all
could get inside quickly for a crash dive. Although a volunteer, I
was not displeased when R.N. officers became available and our
services were dispensed with, particularly since I had already been
accepted for duty in Coastal Motor Boats (CMBs).

4 Belgian Coast and Heligoland Bight Operations

For some time we had heard of the fast motor torpedo boats that the Royal Navy were experimenting with and which had been operating on the Belgian Coast (based at Dunkirk) with the maintenance base and headquarters at Dover. I had applied for this arm. Though it was mainly a Royal Navy affair there were insufficient R.N. officers available to man the boats so R.N.R. and R.N.V.R. officers were taken for command and also as second officers. Most of the second hands were midshipmen R.N.R.

I went direct to Dover and was given a temporary appointment, for experience, as second-in-command to a New Zealand R.N.V.R. officer. After a month operating from Dunkirk at night along the Belgian Coast to Ostend and Zeebrugge, I was sent to the commissioning base at Haslar, Gosport (now H.M.S. *Hornet*), to take over the command of a 55-footer then building at Camper and Nicholson.

CMBs came in two sizes: the larger was 55 ft in length, originally fitted with twin 250 bhp Thornycroft V12 engines, carrying a torpedo and four depth charges or, alternatively, two torpedoes and two depth charges. The smaller 45-ft CMB had a single Thornycroft V12 engine, or a 250 bhp, six cylinder, Fiat aero engine, and carried one torpedo and two depth charges. The CMB hull was a single-step Thornycroft design hydroplane of double skin diagonal mahogany planking. Some Green, V12, 250-300 bhp, aero engines were also used. These boats, according to the engines fitted, had maximum speeds of between 35 and 37 knots. Later, the Thornycroft Y12, developed from the V12 and giving 375 bhp, raised the speed by a further knot or two.

The first time I encountered or identified torsional vibration was with the Green engine, the crankshaft of which vibrated like a tuning fork. With one or two cylinders misfiring the shaft could, and sometimes did, fracture.

I met my first Wren when at the Gosport CMB base. These girls took the place of the traditional Marine servants, who were wanted for much more important duties (generally as guns' crews) than looking after officers' personal and mess needs. Though the Wrens did an excellent job from then onwards, and have long since become an integral part of the Royal Navy, they were new to us in those days and, being mostly good lookers, had a very disturbing effect upon young officers – bringing them their morning tea, laying out uniforms and drawing baths. Considering all things, there was relatively little trouble but some girls left to have babies, the particular officers concerned being posted elsewhere – usually out of the CMB service to other less attractive appointments.

While at Haslar I met some of the flying instructors at Fort Grange (the Gosport School of Special Flying) and took them on one or two CMB trips which they enjoyed. In return, I had an open invitation to fly with them and there I met Colonel Smith-Barry, the Commandant and father of the 'Gosport System' of flying instruction. His system, developed at Fort Grange and taught by picked instructors to others, became the universally accepted method of tuition. Previous early instruction had been a very hit-and-miss affair, depending largely upon the pupil's own ability to master the bare rudiments of piloting – if he survived the process.

Smith-Barry, whom I had taken out in a CMB, listened to the story of my attempts to become a pilot and offered to teach me to fly. So I went to Fort Grange on a number of evenings while awaiting delivery of my new boat from Camper's. The first time I flew with

55 ft Coastal Motor Boat (CMB) by courtesy of J. I. Thornycroft & Co. Ltd.

Smith-Barry, in an Avro 504, he told me I had 'got her' immediately after take-off. When I, at last, had it straight and level he put me through turns and spins. 'SB' made me do the downwind and right-angle approach from the start, *and* the landing, though I don't know how much I was responsible for in the proceedings. Anyway, by the time my boat was commissioned I had reached the stage at which I could have gone solo after that hour or two of Smith-Barry's instruction. He used to tell his instructors to land on the tarmac or apron and ignore the aerodrome – 'Cultivate it', he said.

A somewhat exciting incident occurred while at Haslar when I went along for the ride as a passenger in a CMB taking a senior naval officer from Portsmouth to Lee-on-Solent. Joe Hannen, a well-known yacht designer and sailing man before the War, was also with us. Suddenly, one engine fired back in the carburettor and then caught on fire due to a fuel pipe leak. With the whole engine-room ablaze and the extinguishers quite useless, all we could do was go overboard where we clung to flotation mattresses and the aft torpedo rails. Luckily, we were not carrying a torpedo. Also luckily, we were seen by the seaplane station people at Lee who sent sent out a launch to pick us up.

After commissioning my CMB from Camper and Nicholson and completing torpedo trials in Stokes Bay, I was on passage to Dover when, between Littlehampton and the Brighton light vessel, we came upon a large ship which had just been torpedoed. Being small, we could do little for her. However, reassured by the presence of some minesweepers which were standing by, and having already taken the precaution of setting the depth charges (two for shallow detonation at 50 feet and two for 100 feet), and preparing the torpedo for immediate firing, I circled around to see if there were any signs of the enemy.

We had almost completed a large circle and were preparing to resume our course for Dover when, between the torpedoed ship and the shore, a periscope came up on our starboard side only a few yards away. In fact it hit us a glancing blow, which removed some of our chine at the step and bent the starboard propeller shaft! I had to stop that engine immediately but, turning the boat sharply back on to the estimated course of the enemy, we let go all the depth charges which, because of our relatively low speed, did us nearly as much damage as we possibly inflicted on the U boat. The result of this was that I had to return to Camper and Nicholson for repairs. I claimed a damaged submarine and was eventually conceded a 'possible damage', being commended by the C-in-C Portsmouth.

Eventually, we got to Dover and prepared the boat and equipment for Belgian Coast operations.

To keep these boats in efficient operating trim it was important to see that no extra equipment found its way on board, since any over-weight seriously affected performance. I was often asked, as an engine man, to help a boat captain recover some of his lost 'horses', only to find that the engines were perfectly normal but that he had collected all sorts of equipment such as private store of paint, spare lines and even the odd anchor. The boat was so weighed down that it would hardly get on its step and plane. Always, upon our return to Dover for rest and clean up, the boats were hoisted out and the bottoms cleaned and blackleaded. The zinc plates at the stern, fitted to combat electro-chemical corrosion between the steel propeller shaft and the bronze propellers and stern posts were also renewed.

The CMB base at Dover was under the command of a very tough but able Lieutenant R.N., Welman by name, a stepson of the Royal Navy's top gunnery specialist, Admiral Sir Percy Scott.

On one of my Dover spells I was having tea on board the destroyer *Sikh*, whose First Lieutenant was Victor Crutchley (now Admiral Sir Victor Crutchley, V.C.) with whom I had struck up a friendship. Crutchley was Second-in-Command of the *Vindictive* when that ship, after the St George's Day Zeebrugge raid, was used to block Ostend. He took over command to finish the operation, after Commander Godsal was killed, and received a well-earned V.C. We were walking the deck after tea, when the *Glatton*, a large monitor, blew up. It had been loading ammunition during the day preparatory to proceeding to the Belgian coast to fire on the German lines and submarine shelters. One of the magazines went up, and though we were only a few cables' length from her we did not feel the explosion owing, apparently, to the fact that we were below the level of her main deck and the force of the explosion went directly upward. It was however awe-inspiring to see large pieces of ship's turret, weighing hundreds of tons, thrown into the air, and we watched fascinated to see where they would land! The C-in-C, Admiral Keyes, ordered the harbour to be cleared and the *Glatton* to be torpedoed. This was rather difficult due to the fact that she was of shallow draught and fitted with blisters for the very purpose of defying torpedoes. To sink her, it was necessary to pierce the blister and then try to put another torpedo through the same hole to hit the hull proper. Some torpedoes dived beneath her, luckily going out to sea through the North harbour entrance; but she was eventually sunk and lay for many years mostly submerged in Dover harbour.

On another occasion, while at Dover, I visited the airship station at Kingsnorth and did a patrol in one of the North Sea Star Class airships (Astra Torres type semi-rigids). I had asked to go in the first place, but didn't bargain for a solid eight hours of patrol. It was very boring.

Our Dunkirk operations were intended to hit the Germans at Zeebrugge and, therefore, were mostly night affairs. Our navigating officer or planner of the earlier operations was Erskine Childers, who wrote *The Riddle of Sands*, a book that still holds me. Childers was much older than any of us, but he was first class and with more than average guts, a very interesting man to talk to so long as the Irish question was avoided. He became fanatical on the subject despite the fact that he was not, I believe, Irish born. Most tragically, he was executed as a rebel, in 1921, by the new Irish Free State Government. He remained a rebel when he thought they had sold out to the British Government. The excuse for shooting him was that he had a small pistol on him when arrested. 'Judicial murder' it was called at the time, since it was Michael Collins who had given him the pistol and who himself was assassinated later.

There was a café or bistro in Dunkirk, opposite the Cathedral, called, if I remember rightly, Jules, and here we sometimes repaired with Childers where agents would give him information on German destroyer movements, due into or out of Zeebrugge. Acting on this, Childers would plan an operation for the particular night, either to intercept or go into the harbour and see if we could torpedo any destroyers or submarines still lying alongside the Mole.

On some of these occasions we had the co-operation of the R.A.F., who would appear over Zeebrugge as we were starting our attack. A friend of mine of more recent years, Ronnie Graham (the late Air Vice Marshal Graham, Lord Lieutenant of the County of Bute), was one of the R.A.F. pilots on these particular raids. The idea was that the noise of aircraft engines and bombs would concentrate the enemy's attention skywards. But the Germans soon got wise to this and we were treated to quite a dusting as we came in. Searchlight directed machine guns and other weapons fired on us though we were usually too close for the large calibre guns to be brought to bear. Even with our pink tinted anti-glare searchlight glasses it was difficult to see anything against a direct searchlight beam. After firing our torpedoes we would turn but had to come very close to our targets and the Mole. Then would come grenades or small mortar bombs. These were most disconcerting since their approach could not be heard above the general din and when they went off the effect was disturbing. On one of these occasions I had part of the foredeck of my boat blown off by a near miss. I should add that there was a sand or mud bank in the middle between us and the Mole, so we had to pre-set our torpedoes for a shallow run to the target. Even then, some hit the bank.

One night we were ordered to stand off the harbour entrance at Zeebrugge to intercept some destroyers expected back from patrol.

Sure enough, our information was correct and as we strung ourselves out in line abreast the destroyers were sighted coming in at a
fair speed. I picked my target and, as I thought, fired my torpedo,
when my snotty shouted that it hadn't gone. So we tried to kick and
push it overboard, but in doing this the torpedo motor trip-wire,
attached to the discharge ramhead, started the motor and with the
torpedo tail and propellers just in the water it was slammed back
into the boat. All of this happened very quickly, and as we were
almost upon our target I crossed the enemy's bow and dropped all
depth charges previously set to go off at 50 feet. Air reconnaissance
the next day showed a sunken destroyer in shallow water, but none
of us could claim it individually.

I should explain that the torpedo in a CMB rested in a disposable
cradle on rails in a trough running from just aft of midships to
extension rails a few feet beyond the transom. The forward facing
nose or warhead of the torpedo fitted into a ramhead at the end of a
long piston or ram. The forward end of the ram cylinder had a pipe
connecting it to a gas pressure bottle in the engine compartment of
the boat. The bottle was fitted with a pistol to fire a cordite cartridge,
which was operated by the pulling out of a steel wedge by the captain's torpedo firing lever. We had a spate of launching failures
similar to my own owing to the cardboard cordite cartridge cases
becoming damp and misfiring. Later, brass case cartridges were
issued. Since the torpedo faced forward, the boat itself was aimed
at the target and the torpedo launched backwards; the trip wire on
the ramhead started the motors and the torpedo came ahead, going
into an initial dive of some 19 feet before picking up its pre-set
depth. Though there was a sighting or aiming device provided, to
put on deflection for the target, most of us, particularly at night,
pre-set the torpedo to run at 35 knots and kept the boat at the same
speed, aiming for a collision with the target.

After one such operation, when making back for Dunkirk, I had
both engines stop off Ostend. Some water had got into the fuel
system, presumably from the last tank filling. Dawn came, and while
we were cleaning out pipes, filters and carburetters, there also came
an enemy Brandenburg float seaplane on patrol from its base at
Zeebrugge. The pilot attacked and he and the observer put a
number of rounds into us, injuring one of my mechanics. We manned
our Lewis guns and fired back but did not, apparently, do any
damage. The sequel to this particular affair came some years later,
in Berlin.

On occasion, we carried mines in place of torpedoes for mining
the channel used by enemy vessels entering Zeebrugge. A particular
boob was made one night when we overshot our target area, thinking
we had not run the distance. A marker buoy having a small blue

light, used by the Germans for guiding their destroyers into harbour, was missed or had been moved or extinguished. We carried on until we saw a small vessel with two lights, one above the other. Thinking it to be an enemy ship, we dropped our mines in a line from the lighted side. Returning to Dunkirk and debriefing, we were later told that this was the international gate vessel marking the entrance to Dutch waters, The Hook of Holland and Scheldt! There was considerable diplomatic activity, and apologies, for this unfortunate occurrence.

Another duty we had, one we didn't much like, was making a smoke screen around a monitor which, on occasion, fired on German coastal positions and the submarine shelters at Ghent and Bruges. After a morning's operation we were sometimes invited on board for lunch, particularly if firing was to continue in the afternoon. I was once asked by the gunnery officer if I would like to come to the fire control top and watch the shooting. Leaving my second officer to carry on with the smoke exercise, I accompanied the gunnery officer and found that I liked this better than being in the boat making smoke, because it was quite disconcerting to pass on the shoreward side of the monitor when the main armament went off. At least in the fire control top one knew when the guns were going to fire. The monitor was the *General Wolfe*, which had two 12 inch guns in the forward turret and an enormous 18 inch gun in a kind of super-shield on the after deck, which fired a shell of some 3,000 lb. This gun was taken from the *Furious*, one of Admiral Lord Fisher's lightly armoured cruisers that had been fitted with a hangar and flying deck for aircraft. The Captain, whose quarters were aft underneath the gun, showed me all the bent cabin beams resulting from the recoil. When these guns fired a sheet of flame and smoke was projected about a hundred feet or so from the muzzles and one could actually see the shells, for a brief period, on their way to the target.

The Germans also had a long range gun which fired on Dunkirk. This was in an emplacement at Leugenboom, some ten miles south of Ostend and about twenty-five miles' range from Dunkirk. I think it fired a 9 inch shell.

A lookout on the top of the cathedral could see the flash when the gun fired, and a signal to the lightship in the harbour (brought in from the Dunkirk roads at the start of the War) would start up 'Mournful Mary' – her siren. Then, in about 80/100 seconds, the shell would arrive with a big crump in Dunkirk. Once we saw it make a direct hit on a ship's loading crane on the quayside and bring it down across the vessel. When the Germans were in retreat they wrecked the gun by depressing the muzzle until it was level with the concrete parapet and fired a shell at point blank range.

While most of my leaves were spent at home during spells at Dover from the Belgian coast, I went to Paris on two or three occasions when short leave was possible from Dunkirk; and here I met Elliott White Springs, who was originally one of a number of American volunteers who had joined the Royal Flying Corps as fighter pilots. He edited and completed a fellow countryman's day-to-day diary (*The Diary of an Unknown Aviator*) after the latter was killed. Springs also wrote a number of excellent books on the same theme, giving the ambience at that time of a fighter pilot's daily life, hazards and play. From this meeting I often joined him and others when they were spending their leaves in London, at the Savoy, and these very wild times were well described by him in his books.

*   *   *

During one of my home leaves I met (the late) Lieut. Colonel Harry Delacombe, R.N.A.S., Commandant of the Naval and Military Kite Balloon Headquarters at Roehampton, who invited me there for lunch and a look around. Here they trained kite balloon observers and also budding airship pilots, who first had to do a cross-country free balloon flight as initial qualification for their lighter-than-air ticket.

During lunch, after quite a few drinks, the officer who had been showing me around with Harry said that he had to do a parachute drop in the afternoon and was I interested? Without thinking I said, 'Yes', and regretted it immediately afterwards. Nevertheless, I then had to go through with it and, with the harness fitted, after scrambling into the basket I was attached to a Guardian Angel parachute. This, like all parachutes at that time, was of the static type, the 'chute being folded over a large diameter aluminium dish and lightly held on by thread. The idea was that the dish, attached to the basket, ensured an opening at the parachute mouth of about two feet in diameter. When one jumped the threads were broken and the 'chute dropped a hundred or so feet; thereafter it opened with a bang giving one the feeling, in deceleration, of being hauled backwards. I must say I was nervous, sitting on the edge of the basket ready to go and looking at two or three thousand feet of cable to the ground winch. It was the first time in the air that I had experienced the sensation of height, a sensation I have never enjoyed. However, screwing up my courage and letting go, the sensation of dropping was not particularly marked, and when the parachute had opened it was all quite exhilarating!

Incidentally, Harry Delacome, who died at the great age of 86 in January 1959, was one of the small band of early aviation journalists.

*   *   *

Having completed a number of operations on the Belgian Coast I was transferred to the CMB group at Osea Island, a base in the Blackwater River in Essex. This group operated with the First

Light Cruiser Squadron at Harwich, commanded by Admiral Tyrwhitt. We would leave Osea for Harwich in the afternoon and be hoisted into the davits of the light cruisers. The whole Squadron then steamed through the night towards Heligoland Bight where, at some unholy hour in the morning, and on our side of the Bight, we would be lowered overboard to go over the minefields (thanks to our shallow draught) and see what could be found to torpedo. Mostly it amounted to damaging and sometimes sinking mine-sweepers and patrol vessels. I, personally, only saw one enemy destroyer in the distance and don't think any of these were sunk by us in the Bight, as distinct from our more successful night operations on the Belgian Coast.

The chief danger in the Bight, since we were on daylight or dawn operations, was from aircraft. On one particular occasion (I was not there) the whole flotilla of CMBs was attacked by enemy aircraft and badly shot up, some being sunk or disabled and others drifting into Dutch waters, where the crews were interned in Holland until the end of the War. After such operations we rendez-voused with the Light Cruiser Squadron at an agreed time, were hoisted aboard and brought back to Harwich. The tricky part came when making the journey back to our base at Osea from Harwich, particularly at night since there were no lighted buoys. If the tide was on the ebb one often went aground off the long pier at Walton-on-the Naze or on the mud banks in the Blackwater. Many times I and my second officer were waist deep in cold water heaving the boat off the mud.

Osea Island's only claim to humour was that it was originally the site of a home or curative establishment for well-to-do alcoholics, and when it was being cleaned up to build the CMB base empty whisky and gin bottles were found under every bush and buried in the grounds; the staff having presumably succumbed to bribery by those who could not face the cure.

The Osea boats were mostly of the 45 foot type, a size suitable for carriage by the Light Cruiser Squadron.

Through a mutual friend, Sir Eustace Tennyson d'Eyncourt, the Chief Naval Constructor, I tried to get transferred to the newly created R.A.F., and it was this effort that caused some stir regarding the false age I had given when joining the Service. Since the CMB arm had higher priority than others, I was not allowed to make the transfer but, 'In view of (my) past service, their Lordships decided to overlook the matter of the false age', which had involved the granting of a commission when I was too young to have one!

5 *Back to the Eastern Mediterranean*

In the late Summer or early Autumn of 1918 there came a rumour that Turkey was likely to give up the fight ahead of her allies, the Central Powers, and it was thought that one way of expediting this was to try and force a passage through the Dardanelles. Volunteers were asked for to man some CMBs to do the job, since there was to be no repetition of big ship involvement. We could go at speed over the minefields, past the forts and into the Sea of Marmara, to demonstrate off Constantinople itself and perhaps sink some shipping.

I was appointed captain of a CMB and engineer officer to the twelve, the full complement for the operation. With a lively Canadian, Paul Earl, who had been my second officer on the Tyne and followed me into CMBs, we were given the job of loading the first six boats on board a merchant vessel in the London Docks and accompanying them to Port Said where they were to be taken in the davits of a light cruiser to some suitable Greek island to prepare for the dash through the Narrows. Paul Earl was cheerful, energetic and most capable. With us also came the mechanics for the boats and the shore personnel of artificers (fitters) and torpedomen.

After many delays, and when we were already steaming down the Channel, the Turks gave up their part in the War. We were then ordered into Plymouth but eventually told to proceed to Port Said, though we could glean nothing further of their Lordships' thinking at that stage. However, as we were passing through the Bay of Biscay the whole war came to an end and we had a very dull Armistice Day somewhere off Cape Finisterre – lots of drink but no fun and no real relief. It was interesting to be hailed by surfaced German submarines on their way home; some, by their questions, not being quite certain what had actually happened beyond the orders they had received to stop offensive action.

Arrived at Port Said, we unloaded the boats and were given moorings off the island on the Sinai side of the Canal opposite the town.

There we stayed for more than a month. I then received a signal from the Admiralty ordering us to prepare sufficient stores for a year's operation in the Caspian Sea and to arrange with the Senior British Naval Officer for transport to take us to Batum in the Black Sea.

We had had our difficulties with the Senior British Naval Officer soon after we arrived. First, our boats had to come out of the sea at frequent intervals, to clean and avoid soaking up too much water, and his staff were dilatory in arranging this for us. Second, the S.B.N.O. didn't like us trying out the boats in the harbour or Canal since there were uninformed complaints to him that our wash, of which we had little or none (only spray), was dangerous to to other small craft and also to the Canal banks.

Eventually, the S.B.N.O. agreed with the Canal Company authorities to provide us with a floating crane when we wished to have the boats out for examination and cleaning. Later, I persuaded him that these boats were harmless and took him for a ride in one. After this trip he asked if I would run him down to Tewfik where the French colony and Canal officers were giving a big post-Armistice dinner to all the important personages, naval, military and civil. This we did and I think our record, as it then certainly was, still stands. We did the 90 miles from Port Said to Tewfik in two hours and thirty minutes, averaging 35 knots or over 40 mph.

About a week later, the Chief Engineer of the Canal Company approached me and asked if they could have a demonstration of the CMB's capabilities as a possible means of rapid transit through the Canal to reach trouble spots quickly in cases of emergency. They had an old French torpedo boat hull, the *Aigrette*, converted to a luxury inspection vessel, but she was unable to use her speed potential, because of the wash created, and could hardly exceed 10 knots without causing trouble. So we took some officials down the Canal and they were very impressed with the speed at which we could travel without a damaging wash.

As the result of this demonstration the Canal Company placed an order for two CMB type hydroplanes with Thornycroft, fitted out as inspection vessels. I heard of this after my return from Russia when told, rather sheepishly, by Thornycroft that they couldn't give me any commission on the deal as I was still an officer! I hadn't asked for or expected anything.

Eventually, after a month's delay, we were on board a cargo ship bound for Constantinople (now Istanbul) and Batum. I was also acting as Admiral's messenger as far as Constantinople, with despatches from the C-in-C East Indies Station to the Admiral in charge in the Bosphorus (Sir Gough Calthorpe) whose flagship was H.M.S. *Superb*.

When our motor mechanics (artificers) heard of the impending operations in Russia their C.P.O. ringleader, a former shop steward at Austins, demanded to see the S.B.N.O. to request their return home for demobilization. When I asked this man how long he and the others had been in the Service he replied 'Since January'. When I asked 'Which January?' he replied 'January 1918'. They were the conscripts of the War, and though some were good enough fellows there were the usual disgruntled ones and sea lawyers. I told them that they had only been in the Service a dog watch but I would see the S.B.N.O. and put their request before him – giving me the chance of seeing him alone and warning him of the position. He played up well, saw the whole crowd and told them that if they did not obey orders they would face a charge of mutiny. He was very impressive in going into what could happen if the charge were proved. So, somewhat crestfallen, they returned to quarters.

Upon arrival at Constantinople and after delivering my despatches to the Admiral, we were at a loose end for more than a week as the ship had to discharge some cargo into barges, a slow business, and also take on board some Army equipment for the S. Russian campaign, some of which was notified but not located or immediately available.

My own estimates, in Port Said, of the stores we would need for a year's operation had been largely guesswork. We indented home for some further tools, equipment and spares; but the most difficult reckoning was the quantity of petrol and oil needed, since I did not know what was available at Baku, our eventual destination and Russia's largest oil producing area. I was unable to find out whether or not the wells had been damaged or destroyed, so I took on board the ship at Port Said some 7,000 gallons of aviation gasoline in the usual Middle East packs of two 4 gallon cans in a wooden crate, where the nails were often driven into the tins! Allowing for 25 per cent pilferage and leakage, I thought we had sufficient for twelve months. I was not too far out, fluke as it was, and when we handed over the boats to the Russian Volunteer Navy eight months afterwards, there were left slightly less than 2,000 gallons. In the event, we could have got supplies from Grozny, some 200 miles north of Baku, but we were not to know this at Port Said.

The *Menelaus*, the ship carrying our boats, had this aviation petrol as a deck cargo and it needed considerable persuasion to get the crew to stop smoking completely during the whole trip for the safety of us all. But I caught the scent of tobacco above the reek of petrol several times and could only offer a silent prayer that the crew, in having to climb over the deck cargo to get to their quarters, would have the sense to douse any cigarettes beforehand.

Constantinople was quite good fun and there were some passable restaurants which had been opened by the Germans during the War, complete with the orchestras so beloved by them. But they had little sense of the finesse of things and, as soon as any British officers went in for a meal, the orchestra used to strike up 'Deutschland über Alles', which merely acted as the incentive for us to throw the whole orchestra and their instruments into the street. Eventually, they found this type of truculence or bull-headedness didn't pay.

There was also Bertha, who came from Salonika to open a casino in the Petit Champs – the attractive gardens in the middle of the more European part of Constantinople. Bertha was a Lancashire woman and a very good sort, and Bertha's Bar had been known to all who were in Salonika in the War. Her casino was out of bounds to officers, but a bit of disguise with borrowed civilian jackets from our ship's officers got us a very pleasant evening with good drinks and roulette. No young officer who visited Bertha's and had too much to drink, or who had lost excessively at roulette, went away completely destitute. Bertha saw to that, not only because it paid her to avoid trouble with the military occupation authorities but because she really was a decent woman.

One of the sights to watch in Constantinople was the Sultan going to worship at one of the smaller mosques, with his immediate entourage, officials and escort all in gorgeous uniforms. Later, he moved from the Dolma Batchi palace on the Bosphorus waterfront to another residence, since he complained continually about the near presence of foreign seamen upsetting the women of his harem.

In Constantinople at that time there were a number of Russian Imperial Army officers who had fled the Revolution. They seemed possessed of sufficient funds to live it up on wine, woman and song. General Denikin, leader of the White Russian forces in S. Russia, later sent his Chief of Staff, the much disliked General Romanovsky, to persuade these officers to come and assist him in fighting the Bolsheviks, threatening them with loss of citizenship if they failed to give him their help should he succeed in his task of saving Russia from the Reds. Romanovsky was shot dead.

While in Constantinople I went to the former German military airfield outside the City, since I had heard of aircraft, engines and other stores there, all in good condition, and wished to acquire some Bosch magnetos to replace those of American manufacture (Dixies) which had given a lot of trouble on our Fiat engines. In fact, we had fitted all six boats with Delco coil ignition during the voyage to Port Said – an innovation in those days and only then seen on the Liberty aero engine. I got my German magnetos and removed the American ones but retained the Delco ignition. Thereafter, we experienced no further ignition trouble.

Finally, we left Constantinople and sailed up the Bosphorus for the Black Sea and Batum – to more delays. The boats were put on shore in their cradles together with all the stores such as the torpedoes, depth charges, Lewis guns and ammunition, the fuel and oil. We stayed on board the ship while she was discharging and then went to a billet on shore, the house of a prosperous Georgian merchant, where we were well looked after. The next step was to see the R.T.O. and get a train assembled to take the boats and equipment on the long journey to Baku.

While awaiting all this, our old friend of Dover times, the destroyer *Sikh*, steamed into the harbour. She had been ordered from whatever her other duties were in the Black Sea to complete her stores at Batum and then sail for Trebizond to deal with the remnants of the Turkish garrison there who had, apparently, decided to remain belligerent. They would not recognize the cessation of hostilities.

Victor Crutchley, still her First Lieutenant, invited me to come on the trip. The Turks decided to turn in their arms as soon as the *Sikh* entered harbour, particularly after seeing everyone standing to with the main armament trained on the town.

According to the sailing directions, Trebizond's chief industry was the preparation of porpoise oil and anchovies. It was a miserable place in those winter months of 1919, with little or no food and dead, putrefying, horses and mules in the streets (no one bothering to inter them), and half frozen mud piled shoulder high on the edges of such sidewalks as there were. There was also the risk of being deluged from above by *all* the refuse and slops thrown out of the windows of houses into the streets.

We returned to Batum. The train was, at last, being made up under the able direction of Paul Earl, with a dilapidated coach for ourselves, the mechanics and other boat personnel being installed in covered wagons of the '8 chevaux or 40 hommes' type which they made quite comfortable. At the last moment a Greek doctor, resident in Baku and helping to run a hospital there, requested passage for himself and a Russian princess (all young and attractive women seemed to be princesses in those days) who also lived in Baku. We could not refuse.

So, one evening in January, 1919, we started our 3 week journey of about 450 miles.

The rolling stock was in dreadful condition, and though we had already examined the wagons and axle boxes there were numerous panics such as an axle-box fire on one of the wagons carrying our depth charges and torpedo warheads. Luckily, we had taken the precaution of placing these particular wagons at the rear of the train.

The locomotives were also in a very poor state of maintenance, blowing steam from every joint and gland, and the drivers and firemen were even less reliable since the Georgians, a volatile race, were then politically unsettled. Most of the railway was single track and though there was little traffic movement we sometimes had to wait for hours in a side loop for a train going to Batum. There were the expected delays, the failure to keep a head of steam and the odd mechanical failure, but on two occasions came the disappearance of driver and fireman – stopping off in their part of the country, or defecting, we never knew which. I had to drive the locomotive for about 50 miles into Tiflis (now Tbilisi) the capital of Georgia.

The doctor's stores in our carriage were an embarrassment and one night we all awoke, or nearly didn't, to the smell of ether. One of his tins was leaking and there could have been an explosion if the princess had been awake and smoking one of her innumerable cigarettes. The princess slept in her clothes, as we all did to keep warm, but one evening we came upon her completely naked and sponging herself all over from our cooking tin, which was a four gallon petrol can with the top cut out. Russians never appeared to be embarrassed in this respect, and both sexes generally bathed and swam in the nude at that period.

Having spent a day in Tiflis, and after a decent meal in the best restaurant there (Frascati), we collected another engine driver and fireman and continued towards Baku. At about eleven on the morning of the next day, the train was waved to a halt by an old Caucasian Turk platelayer who, having heard (how, I never knew) of 'important officers' on board, wished to have the honour of entertaining them. We went to his little hut by the side of the track where he had a number of bottles of Caucasian brandy, very good stuff, and also the usual hard-boiled eggs. We sat down, but had to keep bobbing up to toast the Czar, the King and Queen of England, General Thompson (the British O.C. at Baku) and ourselves. Then we were expected to repeat the whole performance back to him. After about an hour of this we staggered back, in a haze, to the train and went on our way.

6 *Caspian Operations*

Upon arrival at Baku we were met by Engineer Commander O'Doherty, a bearded R.N. officer and an Irishman who, I believe, came to Baku in the wake of Dunsterforce from Persia. Paul Earl and I were given accommodation on board an armed merchantman moored at the quayside and somewhat nearer the town proper, while our mechanics and other boat personnel accompanied the train, with the boats, to the old Russian Arsenal further round the harbour.

The plan was to carry the CMBs on the decks of merchantmen. They would be put overboard to give chase and attack when any of the Bolshevik destroyers operating from Astrakhan at the mouth of the Volga were sighted. Armed merchantmen were being prepared to carry about half-a-dozen boats between them. The rest were held in reserve at Baku but were later transferred to Petrovsk some 200 miles north and nearer the likely area of offensive operations. One of these merchantmen, an oil tanker, the *Emily Nobel*, was also fitted with three 6 inch naval guns manned by Royal Marines.

These vessels, the *Emily Nobel* and the *Sergei*, were being fitted at Nobel's yard near the Arsenal, so I transferred myself there and was put up in the house of the Russian manager of the yard. It was, apparently, considered an honour to have a British officer as a guest, low ranking as I was, and they treated me royally.

Being the home of the sturgeon, the Caspian provided plenty of caviare. In fact, it was issued to us, in the absence of much else, as part of the daily ration; but the troops complained that 'This 'ere jam tastes of fish'! Having first tasted caviare in Baku, and consuming quantities every day, I never tired of it and still like caviare when I can have enough of the first quality grey-grain to see and not the small teaspoonful they ceremoniously dish out at £5 a time in restaurants.

At about midnight on the second night of my stay at the manager's house there was a knock on my door and in walked his attractive daughter, a girl of about eighteen. Not quite knowing, as a guest, what to do I suggested that she had better go back to her room, and quickly. But she assured me that her father knew about it and it was *his* wish that she should sleep with me. However, I decided that this sort of thing was not done in one's host's house so I ushered her out, weeping!

It was very cold in Baku in those early months of 1919, and while the carrier ships were being prepared we cleaned up the boats and got them shipshape with the engines tuned up. But before we could test run our torpedoes there was some trouble with the Russian Volunteer Fleet, which was composed of armed merchant ships manned by their normal crews along with some officers and ratings of the former Imperial Navy.

The crews had become unreliable following the efforts of Red agents and their propaganda; on March 1st they took over the ships and, with the help of some officers, left their berths for the middle of the bay where they stood off and threatened to fire on Baku. We of the CMBs were sent out to order the ships back, where the Army, mainly Gurkha troops, were standing by to arrest the crews.

They refused to obey the first order to return and then, after warning them again, two torpedoes were fired at the flagship and one other. Unfortunately, because we had not yet had an opportunity of checking them and doing test runs, both torpedoes floated to the surface short of their target and in full view of all. But just as this anti-climax occurred a depth charge from one of our boats released itself in the choppy water and went off; and this, together with a warning burst of machine-gun fire directed at the flagship's bridge, brought all the ships to heel and back to their berths, where the Gurkhas, with their kukris drawn, took over.

We escorted the ships back and I boarded one to arrest the commanding officer, finding him dead in his cabin with a rifle by his side. He had been shot in the back of the head and the whole of his forehead blown out; the other officers and crew members all said that he feared arrest and had committed suicide. They merely shrugged when it was suggested how difficult it would be for anyone to shoot himself in the back of his head with a revolver let alone a rifle.

Visits to the Tartar Club and the Kursaal formed the lighter entertainment in Baku. The former had a good orchestra and there were concerts, parties and receptions to which the British services were invited. We taught the Russians the foxtrot and they instructed us in a Caucasian dance, the Gopak. Some of us got quite passable at it, always encouraged by our hosts, but we were never as good as the Georgians and Tartars since they put a certain wild enthusiasm into their dances.

The Kursaal was somewhat different, being essentially a music-hall-cum-nightclub. It had a stage and an auditorium with tables and chairs where one could eat and drink. Surrounding this central area, some four or five feet above floor level, were boxes for private parties. It was filled every night, mostly with officer personnel of the Services and also ex-Imperial Russian Army officers who were doing little except walking about in the daytime in their impressive uniforms and drinking too much at nights. When they had sunk too much they used to amuse themselves by coming up to a table and shooting at the bottles and glasses. This got to be dangerous, so our Army authorities, copying the old Wild West, collected all side-arms at the door.

Apart from the mass of oil derricks over thousands of acres, Baku is famous as the trade gateway from the Far East to Europe and also for the original Bluebeard's tower – which had more recently been slightly chipped on one side by a Turkish shell.

Our political officer in Baku was Colonel Pepys Cockerell, a quite outstanding man and a brilliant linguist. He came out at about the same time as ourselves and had not spoken Russian before; within a month he was speaking colloquial Russian. He spoke many other languages and knew the African native languages including, I believe, Bantu with its many clicks of the tongue. He had served in the trenches in the Greco-Turk War and, as a spy, was masquerading as a tobacco merchant in Constantinople during World War I. In addition, he could speak French, German, Italian, Greek, Arabic, Rumanian and, I gathered, Hungarian. Within a month I heard him tell a risqué story in Russian, translating it also into colloquial Rumanian and German.

Cockerell never carried a gun and went about solely with a swagger stick at a time when we were ordered to go about armed and in groups of three. Sometimes we were fired upon by, presumably, Red agents and even got a hand grenade thrown at us while riding in a drosky. Latterly, due to the risk of getting typhus we were not allowed to ride in these vehicles since the upholstery was lice ridden. Bed bugs were the rule rather than the exception, particularly in the CMB carrier ships, and we even got to favour cockroaches which were alleged to keep the bugs at bay. I suffered near blood poisoning from bug bites, and they practically fed on Keating's powder. There was, of course, no DDT in those days.

Some of the high Tartar princes and merchants had very attractive wives but were reputed to prefer boys, or at least male company. A few of our younger officers thought this preference could be used to advantage with the wives, until one snotty had to drop out of a

45 ft CMBs proceeding to arrest mutineering ships of the Russian Volunteer Navy, Baku, March 1919.

45 ft CMBs being transported on the deck of S.S. Sergei, *Caspian Sea.*

high window when chased by an infuriated prince who had returned earlier than anticipated.

At Easter, then celebrated towards the end of March, a number of us went to the cathedral where we held lighted candles for hours, until the ripple of rifle fire outside made us leave quickly – thinking that the revolution had started in Baku. But it was only the Russian way of celebrating their Easter. The Russian New Year was different from that of Western Europe and was changed by the Bolsheviks in 1917 to January after, I believe, the fall of Kerensky's Provisional Government.

In the Spring, with the ice gone from the northern part of the Caspian, the carrier ships sailed to Petrovsk with most of the CMBs, leaving a few at the Baku Arsenal, which was to remain as a repair base. The complement at Baku was reinforced by a Lieutenant Robinson, a Thornycroft man in uniform and a very good engineer officer.

Petrovsk put us within striking distance of any Red destroyers leaving the Volga, and we were further reinforced by some Short seaplanes for spotting purposes and light bombing. Inland from Petrovsk and near Grozny, another oil field area, was a squadron or so of DH9s. Later came some DH9As fitted with the Liberty engine; all the air side was under the command of Wing Commander 'Ginger' Bowhill (to become famous in the Second World War).

When they first received the DH9A the squadron personnel at Grozny were unfamiliar with the Liberty engine and its Delco ignition and I went along, at Bowhill's request, to help. Delco ignition was chosen by the designers of the Liberty, since this was a 45°V, 12 cylinder engine as distinct from the usual 60°V, the smaller angle being decided upon for lower drag installation. Also the Delco system was lighter than magnetos, more readily available and it was more reliable than American magnetos of the time.

The Russian Volunteer Navy at Petrovsk had some French FBA flying boats and I was asked to look at the engines and get them running since these machines had not been used for months. Some had the water-cooled Salmson (Canton-Unné) and others the Monosoupape Gnôme engine. These I had dismantled and, after inspection, reassembled and run up. They were perfectly satisfactory but the flying boats were in poor condition and required re-rigging. I pointed this out to the Russian pilots and showed them the slack in the wires and flying controls, but they were impatient to fly. In fact, they invited me to fly with them early one morning in the Salmson-engined boat but I refused and told them that they were asking for trouble. They persisted and both pilot and observer were killed, the machine diving straight into the water from a few hundred feet after porpoising about the sky with one wing down. One could see the pilot going to extremes with the controls in his endeavour to keep the machine on an even keel.

In our operations north of Petrovsk we were away from the carrier ships and worked from a small creek in the mouth of the Terek River, home country of the Terek Cossacks. During these operations we became very friendly with them and their Ataman, who used to invite us for drinks and food. Wild affairs they were.

Before we left these hospitable people, our C.O., Commander Robinson, V.C., the First Lieutenant, Slayter (an Admiral in the Second World War and now dead), myself and one or two other officers were invited to a feast and were all made honorary Terek Cossaks. This was the toughest of all the feasts we attended in South Russia, and that was saying something. But being very young and resilient we could (just) take it. The party started, at eight o'clock in the evening with zakuski (Russian hors d'oeuvres), which consisted of raw salmon with sultanas (often pressed into one's hand by the large and none too clean hand of one of our many hosts), caviare and the usual hard-boiled eggs, radishes and onions; accompanied by a steady flow of vodka. There were toasts by our hosts to ourselves and by us to them – toasting the Czar, the King and Queen of England, the French President (rather more coolly because of the Odessa incident) and the President of the United States.

When the time came, after dinner and some hours later, for our initiation as honorary Cossacks, each in turn had to stand up, a difficult feat at 3 or 4 a.m., while the Cossacks chanted or sang around us and we drank from a loving cup filled with all the drinks available, turning it upside down over the head to show it was empty. Afterwards, we were allowed to go to our quarters. The feast and ceremony were held in the village hall or *rathaus*, which was built on stilts or wooden columns with the main or first floor above ground level. I vaguely remember easing myself down the stairs and into the night when I tripped on the body of a fellow officer, and there I lay sinking into virtual unconsciousness. Awaking to semi-consciousness some hours later, I found myself back in my quarters with two Cossacks standing guard at the end of my camp bed. Apparently, it was the custom to post two Cossacks to look after each newly inducted Cossack until he came round to a more or less normal state again. I have never since that time eaten and drunk so freely, and it took a day or two of fasting in every respect to recover fully. I suppose most of us had experienced, for the first time, the real effects of alcoholic poisoning.

Before finally leaving the area we gave our Russian friends a return match, where whisky was the basic British offering backed up with vodka and the Caucasian wines and brandy. It was interesting to see the reaction of the Russians to whisky, which they seemed to like. They went out like lights and far more quickly than we did on vodka. The effect of ethyl alcohol upon the human system in the different forms of drink, wines and spirits (grape and grain) is quite marked from one to the other. Vodka seems the safest way of taking a lot of alcohol because, in the process of its making, it is treated with activated charcoal and cleansed of the poisons which contribute to the headache and nausea usually experienced when taking too much of the other kinds of drink. More enlightened Russians sometimes added flavouring to their vodka in the form of orange bitters, made from the peel, or herbs.

We had one or two encounters with destroyers in the Caspian, but nothing particularly decisive. There was 'the battle of Alexandrovsk', a small harbour on the Turkestan side of the Caspian. Having heard of a Red destroyer there we went to investigate. Three CMBs were lowered, went into the harbour and found the destroyer moored to a jetty. After warning the crew through our interpreter to get ashore we put two torpedoes into it.

On another occasion, I was on board the *Emily Nobel* when a couple of Red destroyers were sighted but the weather was bad and there was quite a sea (the Caspian, being shallow, could be whipped up quickly into a bad state) making it difficult to lower our boats. As the ship was being turned to allow those boats that were ready the shelter of a lee while being lowered, the destroyers opened fire with their 5.1 inch guns. Russian naval gunners were good and served their individual guns over a long period. The first salvo from one destroyer straddled us at a range exceeding 5,000 yards. The next hit us, one shell exploding at the base of the engineroom casemate, killing the chief engineer who had come on deck to see what was going on. Our 6 inch armament, manned by Royal Marines, then opened up but seemed to have difficulty in getting the range. Eventually, a shell hit one destroyer in the engineroom and partly disabled it. By the time we had lowered the boats, which because of the sea could not go at any speed, the destroyers had made off and were soon out of range of our deck armament and the boats.

The Caspian was ill-charted and since it was shallowing yearly the carrier ships often found themselves aground in the most unexpected places. Apparently, in the past, those officers of the Imperial Russian Navy who had misbehaved were banished to the Caspian and told to survey and re-chart it. But they mostly stayed in port and did nothing except perhaps drink. We navigated by the local knowledge of our Russian merchant captains and officers and from the odd chart and map, all dated in the 1800s.

One of the less pleasant, in fact most unpleasant, aspects of our stay in the Petrovsk area concerned invitations from the local Russian Army Commander (appropriately named General Popoff) to attend executions of people they had seized and who were alleged to be Red agents and saboteurs but who could have been anyone that the locals had a grudge against. There was a court-martial, but this was a perfunctory affair since they were going to do away with them in any case. We all tried to duck out of these occasions, and eventually resorted to drawing lots for attendance. These unfortunates were either hanged or shot, and they certainly did not know how to hang a person properly. The executions were followed by the Russian idea of a breakfast, which developed into a feast of some hours, but none of us felt very much like eating after witnessing these affairs.

In the late summer, we were ordered to hand over our boats to the Russian Volunteer Navy having first given them a period of training in handling and also instruction on the torpedo. After a week or so of intensive training we were told to get ready for the return home and to leave all stores.

Being used to handling horses a Russian cavalry officer seemed to be able quickly to fly aeroplanes, but in a rather dashing and often reckless way. The Russian Naval officer, without that background, appeared to have the dash but very little judgement and a number of our precious CMBs were pretty badly bashed about within days of being finally handed over.

I was not to get out of South Russia immediately since there was a request from General Denikin's staff for machine gun instructors. As I had passed with high marks in machine gun operation I offered to be an instructor and went to an Army group in the Rostov area. I was instructing, through Russian interpreters, groups of officers and men to be formed into machine gun companies.

Eventually, we were operating in the line. *Line* was, however, a misnomer since there was no line in the normal sense, the fighting being mostly open warfare. We were on the Sea of Azov side of Rostov, on the River Don, where there were small brushes with the Reds who had pushed back the Volunteer Army. Later, we had to retire to Ekaterinodar in the Kuban Cossack country. There, the Kuban Government was undermined by Bolshevik and Green Guard influences – the latter not necessarily being allied to anyone. In fact, the Green Guards were a 'peasant army' antipathetic to Denikin's Volunteer Army and also to the Bolsheviks. They were led by a tough called Mahkuo.

In this period, we underwent the depressing and somewhat frightening experience of being captured and held prisoner for a few days. We had been patrolling on a train of flat cars with our machine guns, not even armoured, in advance of the army group, when we were ambushed by a group of Reds in a village that was thought to be evacuated. It wasn't! Five Russian officers and myself were incarcerated in a windowless wooden hut with an earth floor – and I with dysentery. The next day a Red officer, or Commissar, came and said that they were going to shoot us, which they did in the case of the Russian officers. They took me out too, but apparently had some doubts about a British officer. I was eventually released from 'durance vile' two days later when our army group advanced into the area.

For this interlude I was awarded the Russian Order of St Stanislaus, with Swords. Apparently, General Denikin had the authority to bestow such Imperial decorations.

Churchill wrote of the Russian Civil War in '*The World Crisis – the Aftermath*':

'During the year 1919 there was fought over the whole of Russia a strange war; a war in areas so vast that considerable armies, armies indeed of hundreds of thousands of men, were lost – dispersed, melted, evaporated; a war in which there were no real battles, only raids and affrays and massacres, as the result of which countries as large as England or France changed hands to and fro; a war of flags on the map, of picket lines, of cavalry screens advancing or receding by hundreds of miles without solid cause or durable consequence; a war with little valour and no mercy.

'Whoever could advance found it easy to continue; whoever was forced to retire found it difficult to stop.

'Kolchak first and then Denikin advanced in what were called offensives over enormous territories. As they advanced they spread their lines ever wider and ever thinner. It seemed that they would go on 'till they had scarcely one man to the mile. When the moment came the Bolsheviks lying in the centre, equally feeble but at any rate tending willy-nilly constantly towards compression gave a prick or a punch at this point or that. Thereupon the balloon burst and all the flags moved back and the cities changed hands and found it convenient to change opinions, and horrible vengeances were wreaked on helpless people. A war of few casualties and unnumbered executions!

'Denikin's forces foraged over enormous areas. They boasted a superficial political sway. They lived on the country and by so doing soon alienated the rural population which at first had welcomed them. Had he collected the necessary supplies at one spot in the South for a direct dash to Moscow, and had he seized the psychological moment just before the Siberian armies began to fade away, he would have had a good chance of success. But there never was a thrust; no Napoleon eagle-swoop at the mysterious capital; only the long thin lines wending on ever thinner, weaker and more weary.'

Finally, I was thanked for my services and left for Novorossisk, on the Black Sea, managing to get on board one of our naval sloops bound for Constantinople via Yalta and Sebastopol. I had suffered dysentery for about six weeks without proper medical attention and was in a poor way. Anyone who has had this affliction can imagine the sheer unpleasantness of living for weeks, as I did, in a railway wagon of the '40 hommes or 8 chevaux' type with the lavatory accommodation consisting only of a hole cut in the floor.

The sloop we boarded was filled with refugees, many having to sleep on the deck. Some Russian aristocrats were also on board; among these was Count Tolstoi, a close relation of *the* Tolstoi, and his princess daughter. She was given the First Lieutenant's cabin, with most embarrassing consequences. The No. 1 had not informed her about the heads nor had he, apparently, provided any receptacle for her convenience during the night. We thought he was looking rather glum and embarrassed after landing the Count and his daughter at Yalta, but at midday he cheered up and said it was not often that anyone had his bunk wetted by a real princess!

We proceeded to Sebastopol, where we lay for a few days, going ashore, visiting Balaklava and the site of the Charge of the Light Brigade. There we met General Wrangel, who was eventually to take over command from Denikin for a brief period; but it is now history that the whole White Russian effort fell apart, and when Great Britain and France ceased to give their support there was no further hope for the counter-revolutionary effort.

Arrived at Constantinople I went into hospital for treatment of my dysentery which was a great relief since I had long run out of Dr Collis Browne's Chlorodyne, the universal panacea for this type of complaint in those days.

Finally, I was ordered to Malta to take command of a number of minesweeping trawlers and bring them home to Plymouth. They were coal burners and we had to call at places such as Cadiz, Lisbon, Vigo and Bordeaux to pick up fuel. Eventually, at long last, I was demobilized, after serving for about five and a half years, being only twenty-one years old when I left the Service.

The Author as a Lieutenant R.N.V.R., age 21 (1919).

7 *The Twenties*

Since the War had interrupted my engineering education and training at an important stage, and because I had to earn a living, it was necessary to decide quickly what to do before my war gratuity ran out.

Though I had a good grounding in engineering and some varied experience on internal combustion engines, I also had a leaning towards medicine – I had once helped a young surgeon probationer in the War to operate in a destroyer for appendicitis on a very fat stoker and had also pulled a tooth for a reluctant patient of the same doctor since I had the stronger wrists. But this hardly put one on the road to medicine, though it could have encouraged someone really interested as I was. But having nearly reached the age of 22, with a further seven years needed to qualify as a doctor, I decided to keep to engineering and educate myself by the jobs I could get.

The false age I had assumed at the beginning of the War inferred, to prospective employers, an apprenticeship of nearly five years, which was important at that time, particularly if one had had no formal technical training or engineering degree.

After about two months looking through the Appointments columns of *Engineering* and *The Engineer* I saw, in the former, an insertion by a small firm of naval architects at Newcastle-upon-Tyne who wanted someone experienced in the installation of diesel engines in ships. When I applied for the job I found that the firm was run by two brothers, Scott and John Gunn. Scott, the elder, was a coal exporter and John was a naval architect, whom I had met in the War when we were both in MLs. I told him of my experience but he agreed to try me out.

The firm was more concerned with engines than ship design and specialized in motor ships, advising shipowners on the best or most suitable engine (diesel or semi-diesel) according to size and type of operation. I had never installed an engine in a ship but used my commonsense. In fact, I caused a one-day fitters' strike at the first yard I worked in when I insisted that a diesel had to be very accurately installed and aligned with the propeller shaft. I told them that one could not resort to the practice so common with reciprocating steam engines in the early days and take them on sea trials with sulphur for the bearings and a hose playing on the latter until they wore themselves into line with the tailshaft.

During my time in Newcastle, which included periods in Sunderland and Glasgow, I went to night classes in Newcastle (Armstrong College) to learn something of mechanical engineering, thermodynamics and physics, but I had to miss classes so often that it became almost impossible to maintain continuity and I had largely to depend upon my own reading.

I travelled widely for this firm, mainly as guarantee engineer. That is to say I was the engineer recommended and approved by the engine builder and responsible for the good working of the engines for, say, the first six months' operational guarantee of a new installation. To qualify, I went to the engine builders concerned such as Sulzer at Winterthur, Switzerland, and Burmeister and Wain, Copenhagen.

Diesel and semi-diesel engined ships gave much hard work to their engineers and though I went to many interesting places around the world, I saw little of them because I was always removing cylinder heads, grinding in valves, pulling pistons or bedding in bearings. All one had time for was a brief dash ashore for a drink and perhaps a meal. But it was good experience and taught me a lot.

At one time, we were installing semi-diesels in some old iron built, three-masted, sailing vessels which, in the past, had done the Australian wool run. The sails were gone and the owner, one of the small boys in shipping, often former shipping managers or clerks who had been quick to cash in on the high freight rates during the War in taking coal to France, thought to make money in the Continental and Baltic trade with the economical oil engine.

Unfortunately, at that time, the more optimistic diesel engine manufacturers, to make a sale often recommended engines too small for the job, on the basis, they thought, that the diesel could be operated at full bore or at least at a high percentage of its maximum power. The result of this was a spate of unreliability which could only be cured by down-rating the engine. This of course seriously reduced the performance of the ship and put up running costs which reacted against the shipowners; so it was with one of our converted iron built three-masters.

The following story will suffice to illustrate what could happen in those days and it contributed towards my final decision to give up marine engineering for aviation, which I wanted to be in anyway.

One day in the early Summer of 1921 we set sail in ballast from the

Tyne bound for Boulogne to load Army stores from dumps left over from the 1914–18 War, and some bought by agents of the Bolshevik Government to help feed the Russian people. This cargo consisted mainly of tinned bully beef, Maconachie rations and pork and beans, etc. The old sailing vessel originally on the Australian wool run had, in her new guise, two 220-bhp Petter semi-diesels installed, giving just sufficient power for a bare seven knots at maximum power.

On our way from the Tyne to Boulogne, the port engine developed overheated bearings (a main and big end) and had to be stopped. I worked through the night, almost single handed, to remove, scrape and bed in both bearings and refit them. None of the engine-room staff, who were ex-steam engineers, had previously dealt with diesels or semi-diesels. It took me twelve hours and I was almost delirious, 'seeing' people, whom I knew were at home, coming into the engineroom, and talking to them! We proceeded on our way but, after a few hours, when I was resting on my bunk, I sensed or heard the engine beat alter and then one engine stop. I leapt out and went to the engineroom and, peering through the top grating, saw the supernumerary engineer asleep. He had allowed the daily service fuel tanks to empty. These were gravity tanks which directly fed the engines and were kept filled by pumping from the main double-bottom tank.

To be asleep on watch is the most heinous crime in a ship. I spotted an open can of oil on the top grating and poured it over the sleeper and then dropped the can on him. He woke up and, seeing me looking at him (I had descended into the engineroom), realized what I had done. He went for me, and while I got knocked about in the ensuing fight I was so angry that I got him on the floor and went on hitting him until he had had enough. Then I logged him for being asleep on watch.

Eventually, we arrived at Boulogne and loaded ship. We left for Leningrad by way of the Kiel Canal which had only recently been re-opened. During the trip to Brunsbüttel, the North Sea end of the Canal, we found that our Captain to whom we had entrusted the purchase of stores for the catering and messing of the officers and crew, had stocked himself up with whisky instead – leaving us only the BOT emergency rations of pickled pork (in barrels) and biscuits, the latter having a high hardness value. There were some tins of corned beef, in addition.

The Captain had started heavy drinking when we left the Tyne and was drunk the whole week in Boulogne. He proved incapable of navigating properly and the first mate, a capable but nervous and now very frightened man, was not allowed by the Captain to interfere with navigation. In fact, we were so far off course as to nearly ram Heligoland. He, the mate, came to me to ask what I thought he should do and we both went to the Captain and tried to persuade him to go to his cabin and lie down, but he would have none of it.

A German pilot boarded at Brunsbüttel to take us to the half-way point in the Canal, where another pilot took over for the last leg to Kiel. The pilots, like most Germans at that time, had been at near starvation level and accepted the Captain's whisky gratefully; but each in their turn became as drunk as the Captain and were quite useless. So, with the Captain's refusal to hand over to the mate, the latter shrugged his shoulders in despair and went forward to prepare for our arrival at Kiel. And I, by now more than somewhat apprehensive, decided to station my second engineer as near as possible to the engineroom telegraphs on the bridge (it was dark) so that he could log the orders given, with their times; since I could not in the engineroom anticipate or disobey a telegraphed order however wrong it might prove to be. Sure enough, when we entered the lock, the Captain rang down 'half ahead' instead of 'slow astern'. There were shouts on deck and on the quayside, then a bump that threw some of us to the engineroom floor. After a pause, the telegraph rang 'full astern' which, after some seconds, resulted in a heavy grinding and the riding of the ship against the quayside. Luckily, the wire 'springs' held. After this, all hell broke loose. The Captain accused me of executing his orders wrongly, and I countered by showing him the second engineer's pencil log of the bridge orders. He, the Captain, was very near collapse and so drunk that he had to be helped from the bridge.

The mate, stiffened somewhat by this episode, helped me get the Captain to his cabin, where we locked him in, telling the Germans that he was unwell. He had DTs and was howling and barking like a dog. The mate then assumed command and took the ship from Kiel into the Baltic.

The course was set for Cronstadt, but here again we got into trouble since the Captain had not disclosed the fact that he had received specific sailing instructions from the Russians before leaving Boulogne. These were for us to proceed to Reval (now Tallinn) to pick up a pilot there to guide us through the minefields covering the approaches to Cronstadt. When practically on the edge of a minefield, we were 'arrested' by a Bolshevik minesweeper and a submarine and escorted into Cronstadt harbour.

The head Commissar of this fortress-port guarding the entrance of the Canal to Leningrad came on board, accompanied by hordes of officials. I found out later that he had sailed on the Cardiff-Philadelphia run in coal boats before the War and, therefore, spoke fairly good English. They seemed pleased to see us when they learned that we were carrying provisions. He and his colleagues were entertained by the Captain, who had recovered somewhat, and they got sufficiently drunk for us to decide to keep them on board for the night to save them possible trouble when they returned ashore.

Leningrad had only just been opened again as a port and we and a German ship were the first cargo vessels to arrive there since the start of the Revolution.

The ship was discharged under armed guard; when some of the dock labour were caught eating from the tinned cargo they were marched off not to be seen again. We were given shore passes, printed on the back of which were instructions to any Russian to whom we showed them to guide us back to the ship. The head stevedores were English-speaking (American-speaking to be accurate) and one invaded my cabin when I was shaving one morning saying, 'Gee, Chief, we want you and your boys to play us at football'. Having fairly recently returned from naval, and military, operations in Russia, I was not feeling all that friendly and told him to knock on the door before coming into my cabin. My shore pass was held up after this for about twenty-four hours.

We visited all the sights and palaces of this still beautiful but then dilapidated city, including the Hermitage and St Isaac's Cathedral. In the latter were a number of people praying, despite the Government's strictures on religion.

We had been cabled by our owner when loading at Boulogne that he would send instructions where we were to pick up the next cargo for our return trip, but none were forthcoming in Leningrad and we were told that they would probably be in Moscow and would have to be collected there. None of the ship's officers were eager to go so I volunteered, setting off with some iron rations and accompanied by a Commissar. The train journey, which used to take about eight hours, took over twenty-four. We were a full ten days in Moscow before we could get any news, though the cable had in fact arrived a fortnight previously. It informed us that we were to proceed to Kotka, in Finland, to load pit props for the Regents Canal Dock in the London River.

The most frightening and at the same time pathetic sights I saw in Moscow, shown to me by one of the foreign newspaper correspondents, were the groups of starving and wild children, some war orphans and all abandoned, who were to be seen scavenging for something to eat. They had apparently devoured all the cats and dogs, and it was alleged that humans were not safe if caught by them. The police kept them away from the main squares and streets.

Back in Leningrad, we were entertained as saviours of the

Russian people and were taken to the Opera, occupying the Czar's box. We saw fairly good ballet and listened to two excellent concerts.

The Nevsky Prospekt (the Piccadilly of Leningrad) was hardly as I saw it before the War. The paving blocks had all been torn up, probably for fuel. Everything was dilapidated and all the bridges over the Neva and canals were rusty.

The Russians were literally at starvation point, and there were soup kitchens serving meals to the hungry populace. The soup appeared to consist mainly of water and cabbage leaves to which had been added salted herrings, which the Russians used to buy in quantity from us before the War.

Arriving at Kotka, the ship was loaded at anchor from barges by women. They packed the pit props in the holds as close as matchsticks in a box and then piled on more as deck cargo.

Kotka used to be Russian Finland, and in the old days the Czar had a summer fishing palace there. The Finns hated the Russians and feared the effects of the Bolshevik Revolution, so we were at first regarded with considerable suspicion when they learned that we had come from Leningrad.

There were literally hundreds of small islands at the entrance to Kotka and in the anchorage. They were really rocky islets, and our crews were greatly diverted from their duties since the Finns, like the Russians, bathed naked. Every day there were a number of unclothed girls swimming around these islets or sitting at the water's edge only a few cables' length from the ship.

Finland was 'dry' at that time and no alcohol was allowed, but there was considerable smuggling by Estonian fishermen who brought in potato spirit. I had made friends with Kotka's Chief of Police, who was a keen yachtsman. One day, after sailing in his small boat, he took me to the yacht club for dinner and offered me the choice of almost any brand of whisky I wished. When I queried this rather unusual gesture for a head of the police force, he said it was contraband and, with a smile ' – surely it was a pity to waste it'.

The Finns could be tough and on the one or two occasions when we went to a smoky type of seamen's hang-out where we could get something to eat and also illicit alcohol, we saw Finns with a load of this fire water on board fighting with flick knives. The resulting melée was something to be seen and though they didn't seem to kill themselves, they were often hospital cases.

Upon arrival at Kotka I had cabled my firm about the Captain. The owner sent out his ship superintendent to pay the Captain off and send him home, the superintendent taking over.

We sailed from Kotka in a high wind with a list of 12 degrees to port due to the deck cargo. As the pilot was taking us round a small island the ship straightened up in the turn and fell over to starboard, assuming a list of 25 degrees. This was due to the fact that the double-bottom fuel tank was not centrally divided and the fuel sloshed over to the listing side.

Upon arrival in Kiel harbour we had to anchor for a day or so while the port authorities decided whether or not to let us through the Canal with that degree of list. Eventually, we were told that if the list were reduced to under 20 degrees we could proceed. The crew then got busy and stacked all deck cargo possible on the port side, which brought the list back to 18 degrees. We were then cleared to go through the Canal but with two helmsmen at the wheel.

Our troubles continued and it took six days to drag ourselves across the North Sea to the London River and Regents Canal Dock.

Before paying off I had to go to the Tower Bridge Police Court with the supernumerary engineer, who was fined £25 with the alternative of imprisonment.

In 1923 I was in Bremerhaven, discharging ship, when the Reichmark was devalued. It was a most harrowing experience, to see the peoples' earnings and savings literally disappearing in days and even hours. We went ashore on the second day of devaluation for a meal in a restaurant and they could hardly give us enough change for £1 Sterling after paying the bill for two. On the same day, three of us had a whole evening in a beer hall on ten shillings. Paper Marks were being printed by the millions and for many millions on an individual note. The Mark, having no value and now in astronomical figures, led the various town councils to print their own notes, equally unbacked, to keep local trading going when notes from the Reichsbank didn't come along quickly enough. These were called Stadt Geld.

Finally, it was Dr Horace Greeley Hjalmer Schacht, who was taken from the National Bank für Deutschland (later joined with the Darmstäder Bank) and made Commissioner for National Currency on the 13th November 1923. The Mark, which had in 1914 been worth DM4.2 = $1 U.S., when Schacht took control had reached DM4,002,000,000,000. He introduced the Rentenmark which, in theory, was equal to the gold Mark but covered by a cautionary-mortgage of the whole of the German landed property and could be exchanged for a like sum in mortgage bonds only. The peacetime rate of the Mark was restored by knocking off the billion sign. Shortly afterwards, Schacht was made President of the Reichsbank by President Ebert; but when Hitler came to power he left in disagreement with the Nazi's financing plans – to be restored as German Minister of Finance later.

*   *   *

I liked the North Country and, in particular, the N.E. Coast where I had this, my first post-war job. The people were tough but friendly. In those days, with little money (£250 p.a., later increased to £350), one could just about get by, but with not even a motorcycle.

My first ship installation was at Jarrow and I had lodgings in South Shields, over a chemist's shop. These were quite comfortable until the chemist's lease ran out while I was there and some fried fish people took over. I was then forced to leave since the scent was all-pervading and, also, my room looked out over the backyard where they prepared the fish.

I then got better rooms on the north side of the Tyne, in Culler-coats. The landlady was a widow and I her only lodger. She was a vinegar drinking addict, the first and only one I have ever met; but the real problem for me was that she had a daughter and was wanting to get her married off. The daughter was a very nice girl but not my type so, after a month or so, I moved to another lodging in Whitley Bay. It was from there I met a number of young people of about my own age who had also been in the War but all in the Army. Some had been badly wounded or severely gassed; one, George Langlands, was in an accountant's office for training prior to taking his Chartered Accountant's examination; one, Charles Read, an engineer in Merz & McLellan, the well-known electrical consulting engineers; one, B. F. Roycroft, was in the Bank of England (Newcastle branch). The others were salesmen of one sort or another.

George Langlands, a Scot, had had a hip joint shot away and lain for some days in a shell hole before they got to him. There were no antibiotics in the 1914–1918 War and it was only by a near miracle that he didn't die or at least lose his leg. But the sheer persistence and interest of a London surgeon saved both his life and leg, though he was in hospital for nearly two years. The treatment of such wounds then was mainly to irrigate them frequently, indeed almost continuously, with Carrell-Dakin solution (of sodium chloride and sodium bicarbonate).

George and his only sister, Constance (Con), lived with their parents. Their father, Robert Langlands, was a ship surveyor in Lloyd's Newcastle office.

Dancing was the done thing in the 'twenties, and we all used to go at least once a week to the various hotels and dance places. I also took up golf, which I have played on and off ever since but at which I perform very badly indeed. In fact, I play at golf rather than play it since I dislike going for walks and golf is a most pleasant way of exercising.

The elder Langlands had a rented cottage for all the year round at Lochranza on the Isle of Arran, an island off the Clyde and between it and the Mull of Kintyre. They went there frequently, at Christmas, Easter and, of course, in the Summer. The family invited me on a number of occasions for my holidays, and I enjoyed these more than any I ever remembered.

Shortly after one of these vacations, I think in 1923, I asked Con if she would marry me and we became engaged; and in 1925 were married in Whitley Bay. I can only say that I have had the nicest and most loyal wife and good companion for over fifty years. She has had to put up with a lot, not the least with my travelling and long trips away from home. We have two daughters, Christine, married with three children, a boy and two girls, and the younger, Annrobin, who has made her own career as an artist and skilled enameller.

Con looked after the two girls all through the Second World War; and by knowing that the family were comparatively safe, I could better concentrate on my work.

8 *From Sea to Air*

I was interested in ships and their engines but it became clear to me that I was never likely to earn any real money in marine engineering, though money as such has never been my prime aim. I did, however, want to work with the aviation engine and be concerned with its development.

In this post-war period, senior ship and engineer superintendents of some fairly large N.E. Coast ship-owning companies, who got their shore appointments after twenty years or more at sea, were being paid only about £450 a year. This often led to the taking of bribes from the dry dock and repair people when a ship was due for a survey and overhaul. The dock touts would approach the superintendents and offer them a rake-off if their estimates were accepted and the ship came to their dock. Naturally, the shipowner, who was the basic cause of this state of affairs, was faced in the end with a much higher account.

Some of the more shady and smaller shipowners, mainly of weekly found vessels (meaning the crew was taken on and paid off after each round trip), were thoroughly dishonest and would get one of their poorly paid captains to log that his vessel had run aground in shoal water (at an approach to a harbour) and, with the engines working to get her off, the hull had become strained causing the engine bearings to overheat. They would then obtain an overhaul from Lloyds, either free or at a proportion of the full cost.

I had personal experience of these tactics when acting as engineer superintendent to a small coasting company on behalf of my firm. I was asked to see one of the vessels in dock after an alleged grounding. It was on a Sunday and I arrived at the dry dock very early, already having some suspicions, and found a gang jacking up the bottom plates to simulate grounding damage. Needless to say, the owner had part interest in the dry dock company. The captain superintendent came along later and both he and the ship's captain asked me to sign the log that the bottom was damaged and the engines needed dismantling as a result. I refused to do this, which led to an almighty row with the owner and the loss of his patronage of my company. But my employer, John Gunn, backed me up.

Finally, I decided seriously to look for a job in aviation; and in an issue of the same journal that brought me this, my first position, an advertisement appeared for an engineer having diesel experience to convert a large petrol airship engine to the diesel or compression ignition cycle. I applied for and got the job as chief experimental engineer, a high-sounding title but one only carrying a salary of £100 p.a. more than I was getting. I felt, however, this would provide the experience I needed and give me an entrée into aviation and contact with the Air Ministry.

The firm, Peter Hooker Ltd., of Walthamstow, was also known as The British Gnôme and Le Rhône Engine Co., since they had taken a licence for the manufacture of Gnôme and Le Rhône rotary aero engines as early as 1913. They were high quality general engineers and also built attachments to printing presses for the printing of large colour advertisements, which was very accurate work. In my time there, they designed, developed and manufactured an automatic bread-wrapping machine for Hovis. Dr Kapitza, the Russian physicist at the Cavendish Laboratory in the 'twenties (who later had his passport confiscated by Stalin when on a visit to Russia), came to us with some rough sketches of a switch to break 1,000,000 volts in 100,000th of a second, for Rutherford's atomic work. We designed it from Kapitza's sketches and with his help made it in our Special Department. It is now, I believe, in the museum of the Cavendish Laboratory at Cambridge.

So, it will be appreciated that Hooker's had considerable mechanical design ingenuity and manufacturing ability.

Peter Hooker had built the most reliable of the Gnôme engines in the 1914–18 War, mainly through their method of making the famous obturator ring by a special tool for rolling the L section. Their Gnôme engines would go for 80 hours or more before needing stripping and overhauling, compared with the 25 hours or so of the French built engines.

They also developed a process of producing forged Y alloy (originally, but I believe wrongly, called Rosenhain's metal). This was a copper bearing light alloy of high strength developed by the National Physical Laboratory, being used, originally in cast form, for pistons and air-cooled cylinder heads.[1] In the forged state we used it for the pistons of the Monosoupape Gnôme engine, which were then fitted with conventional rings entirely replacing the cast iron piston and obturator ring.

The late Frank Halford used forged Y-alloy connecting rods in his $1\frac{1}{2}$ litre racing car engines and for the Gipsy engines he designed for deHavilland. It was, of course, excellent piston material but I never thought it offered any advantages for connecting rods. To

---

[1] The composition of Y alloy was, 4% copper, 2% nickel, 1.5% magnesium, the balance aluminium.

provide stiffness, big end rigidity and strength they were clumsy, fat affairs compared with their high tensile steel counterparts. Consequently they could be debited with a certain amount of 'windage' and oil drag and, as a result, power absorption in the enclosed volume of a crankcase.

The engine I had joined the firm to develop and then convert to the compression-ignition cycle was mainly interesting on account of its sheer size. Appropriately named 'Stromboli', it was a vertical, 6 cylinder job of 12 in bore and 16 in stroke, rated at 1,500 bhp at 900 rpm, and intended to drive a 20 ft diameter propeller directly from the crankshaft. In place of the actual propeller on the test bed there was a flywheel of about the same mass moment of inertia.

There was also a single cylinder experimental test unit of 12 in × 12 in bore and stroke, for performance testing and endurance running. This ran at higher revolutions than the main engine to give the same piston speed.

I eventually got both the engine and unit developed to give near to their rated performance on petrol (gasoline) and then started to convert the single cylinder version to compression ignition. I was helped in all this work by a capable assistant and good engineer, Reggie Schlotel. Hardly any real running and no endurance testing had been done on the Stromboli before I came, since the crankcase,

1,500 bhp E.L.S. Stromboli airship engine.

an Alpax (high silicon/aluminium alloy) casting, had fractured a number of times. It was not difficult to see why. The crankcase scantlings were no greater than those of the 250-bhp Fiat aero engine on which I had so much experience in Coastal Motor Boats. In addition, there were large lightening holes cast in the bearing webs! In one run of only a few minutes at 60 per cent power, all the bearings, except those at each end, were swinging free from their webs and, as the late D. R. Pye put it, 'were hanging like elegant bracelets on the shaft'.

Hooker's was the overspill from Farnborough in those days, where an entirely new and experimental engine could be worked on without the R.A.E. coming directly into the picture and thus being accused of taking the bread and butter from the mouth(s) of the Industry. There had been in the War criticism of the (then) Royal Aircraft Factory because it engaged in the design and manufacture of aeroplanes and engines, this activity being considered the rightful job of private industry. Parliament came into the act, through an M.P., Pemberton Billing, and it was eventually ruled that Farnborough's rôle should be confined to research only, on airframes, engines and allied equipment.

For instance, the 14-cylinder, two-row, air-cooled radial engine, the RAF8, designed at Farnborough, went to Siddeley Deasy (later, Armstrong Siddeley) together with some staff, notably Major F. M. Green and S. D. Heron, where it was developed and produced as the Jaguar.

The Stromboli, however, came after the War and was entirely supported by the Air Ministry.

The E.L.S. Company (the initials were those of the designer of the engine, Ettore Lanzerotti Spina) to whom the contract was awarded by the Air Ministry, was responsible for the Stromboli's design. Peter Hooker, in turn, then received an order from E.L.S. for the development and manufacture of the engine. Hedley Thomson, the general manager of Peter Hooker, was also a director of E.L.S.

Lanzerotti, an Italian, was a charming man with a most dignified presence. He had a neatly trimmed beard and was heavily built. Like many of his countrymen, he was a designer who could produce a neat, clean and tidy engine, rather in the same way as Marc Birkigt of Hispano. Lanzerotti was a very ingenious mechanical designer but a bit weak on stressing and thermodynamics. The Stromboli's design was centred around his (Lanzerotti's) valve gear, a very clever mechanical arrangement. There were eight valves, four inlets and four exhausts, disposed around the periphery of the (detachable) cast iron cylinder head, with their stems converging towards the centre of the head, where each group of four valves was operated by a push rod and rocker arm bearing on a

Single Cylinder Stromboli test unit of 12 in stroke and 12 in bore. It developed 250 bhp.

sliding sleeve carrying rollers. These sleeves were mounted concentrically on a fixed post or pedestal on the cylinder head, one sleeve sliding on the outside of the pedestal and the other in its bore.

It made for a horrid shape of cylinder head, being convex on the combustion chamber side. In fact, at 5:1 compression ratio and with 80 per cent petrol 20 per cent benzole mixture of about 74 octane (meeting the Air Ministry specification of that time) detonation and then almost immediate pre-ignition occurred, which lifted the cylinder head causing flame to issue from the head joint. This in turn blew the water jacket, since the head, when bolted down, clamped a protective steel ring that compressed a rubber ring forming the main water jacket seal and joint.

On one occasion, when running at near full power on the single cylinder unit (230 bhp) the cylinder fractured at its base and, literally, went into the air like a 12 in shell. At another time, when completing a power curve run up to maximum power on the full-scale engine, I saw the flywheel suddenly describe an elipse. Quick as I was to ease the throttle, the wheel twisted from the crankshaft and finished resting on the test bed.

It was a somewhat dangerous and useless engine on its original crankcase, so the decision was made to build one in steel. This was of 80-ton material having seven diaphragms or bulkheads, one between each cylinder to carry the bearings and take the working loads, a top plate or deck connecting them together, bored to receive the cylinders, with steel side plates bolted to the bulkheads and top plate to form a rigid box. All except the side plates were milled from solid slabs of over two inches in thickness, the flange width, down to a web and diaphragm thickness of a few millimetres. Practically every milling machine in the shops had to be reconditioned after this effort!

The engine was, however, only to run twice up to near full power (just short of 1,500 bhp) on the new crankcase since the Air Ministry stopped all contracts. This was before the loss of the airship R101, whose Beardmore engines the Stromboli was intended to replace. Farnborough (R.A.E.) also built a crankcase for the engine, designed on the Warren truss principle with high tensile steel tubes taking the principal loads. This was delivered but not used.

I managed to get the single-cylinder unit 'dieselized' but only completed a few runs before this too was closed down.

It was in the testing of the Stromboli and its single cylinder unit that I first used the Farnborough Indicator, to get an idea of the cylinder pressures and the engine's indicated performance from which to estimate the mechanical efficiency. It was possible to fit the indicating diaphragm unit in one of the four sparking plug holes. Then a lot of 'damned dots' could be seen on the diagrams taken

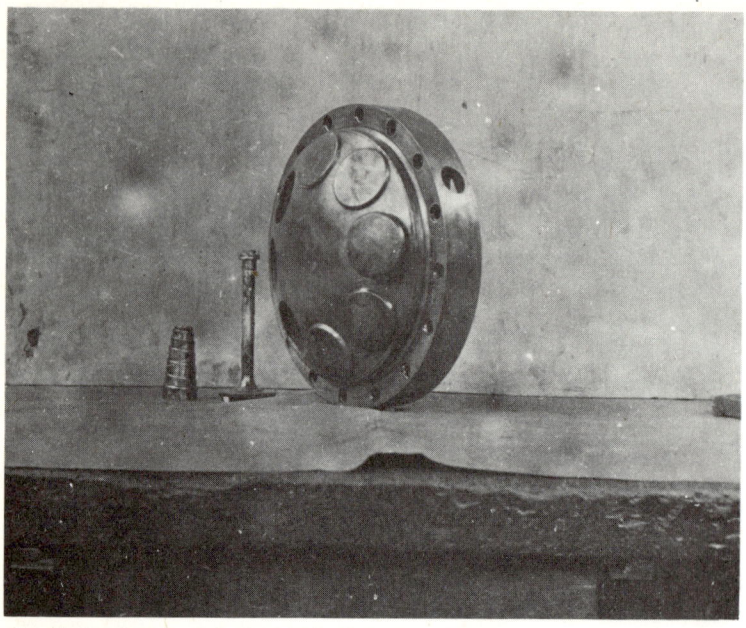

Stromboli eight valve Cylinder Head.

Experimental warren-truss crankcase for Stromboli airship engine, designed and manufactured by the R.A.E., Farnborough.

under relatively moderate power conditions. These showed peak pressures well in excess of normal due obviously to incipient detonation or pre-ignition.

I was also one of the first in the U.K. to try tetraethyl lead. The R.A.E. had made some on a laboratory scale and later received some from the U.S. At my request, sufficient was sent me to try in the Stromboli single-cylinder unit. The result proved quite dramatic and was the only occasion on which I had been able to run the engine completely free of knocking and pre-ignition. It even gave some increase in power showing that detonation, or some similar type of combustion disorder, had previously been the limiting factor. This prompted me to design a piston with a curved pent roof, fitting the cylinder head bulge and, in effect, splitting the combustion chamber in two, which was quite practical with four sparking plugs per cylinder. It also would have given good turbulence. Unfortunately, by the time we were ready to try it in the single-cylinder unit the whole contract was cancelled and everything shut down.

Our General Manager, Hedley Thomson, had a good brain but, like a number of really brilliant people, often went off at a tangent. He thought Y-alloy, which was largely his baby, was *the* universal material and could be used for most parts of an engine. For instance, he decided to try a Gnôme engine cylinder in this material and asked me what I thought. When I said it would just fly off he became impatient, got out his 20-in slide rule and told me that the working stress was well within the limits of the material. I tried to tell him, as mildly as I could, that since the cylinders of the Gnôme were clamped between the two halves of the crankcase, this fixed, and limited, the dimensions of the Y-alloy cylinder to those of the steel one. Also, the bending stress imposed by the exhaust valve push-rod upon the cylinder barrel should not be ignored. He wouldn't listen, however, and put an order on the Special Department (the experimental manufacturing section) to make a Y-alloy cylinder. This was done and fitted to an engine, which was then mounted on the usual Gnôme 'gun carriage' test bed. I advised the tester not to put the engine in the usual escargot cowling, for cooling, since the cylinder would soon come off. It was sufficient, I said, to try it out in the open with its club fan brake, but to keep clear. Sure enough, when the Y-alloy cylinder fired, before the engine had completed a revolution, off came the cylinder, sailing about 70 feet in the air over the Stromboli test house.

While we heard no more of Y-alloy cylinders, Thomson then decided it would be good material for piston rings and asked me to try some in the Stromboli cylinder. I protested but he insisted and I was able to show him later a number of badly worn and feathered rings after an hour or two of running in the single-cylinder unit.

Hedley Thomson had been with Arrol-Johnston in the War and was, I believe, their works manager. Also at Arrol, were Frank Halford and Brodie, who were to be together for many years subsequently in Halford's independent engine design company and, later, the de Havilland Engine Company. Dr Eric Moult was at Hooker's in my time and later joined Halford to become one of this country's capable aero engine designers. Halford thought up the original B.H.P. (Beardmore, Halford and Pullinger) engine, which was eventually manufactured by Siddeley-Deasy (Armstrong Siddeley) as the Puma, a vertical, 6 cylinder, water-cooled motor of about 250 bhp. Thomson had worked out a system of gauge limits, known as the 'Thomson Limits', and these were later developed by the Newall Gauge Co., incorporated in Peter Hooker, Thomson being their manager as well as General Manager of Hooker's.

When I joined Peter Hooker it was in voluntary liquidation, but I took the risk since I was advised that it would last for about three years and, in this time, I could get to know the Air Ministry people and others in aviation, which would be well worthwhile. It was in fact just over three years after I had joined that the firm finally closed its doors.

I saw the rough end of the General Strike at Walthamstow. Since we were in voluntary liquidation and had no bank credit I, with other senior staff, kept piston production going for Gnôme and Armstrong Siddeley engines – incidentally with less than the normal amount of scrap and a greater output rate. Walthamstow was the home of a number of racecourse and other gangs at that time, who exacted tribute from the stallholders in the long street (I think Hoe Street) running more or less parallel to the railway. If the owners didn't pay, their stalls were turned over or destroyed. Of course, the victims were too frightened to say who had done it, so the police could never get any witnesses for the prosecution.

The gangs took full advantage of the General Strike to cause mischief. One day the local chief inspector of police told us they had information that a gang was going to rough up the passengers from the evening trains driven by volunteer drivers. He said that if we wanted to see some action we should keep clear but be near at hand. After letting the gang show its hand, brandishing bicycle chains and throwing potatoes with embedded razor blades, the police charged them with drawn truncheons – not seen much in these days. I have never seen people hit so skilfully and hard without being killed. It took a number of ambulances to cart the injured toughs to hospital.

Peter Hooker was finally wound up in the latter part of 1927, Achille Serre, the dry cleaning firm, buying the buildings and Alfred Herbert the machine tools. Just prior to this Armstrong

Siddeley had received the largest post-war order from the Air Ministry for the Jaguar engine and we were to supply the pistons, which would run into some thousands.

John Siddeley (later to become the first Lord Kenilworth) was put on the spot, since no other satisfactory piston had been tested and approved. But Devereux, our Works Manager, decided that this was a good opportunity and business worth having and developing for the future. He approached Siddeley and made a proposal to set up a forge and machine shop to meet and complete the piston order. Siddeley lent him funds and Devereux bought back the hammers, stamps and tools from Alfred Herbert. Working day and night, seven days a week, he set to work at Slough with ex-members of Peter Hooker's as his senior staff and works personnel. This was the start of High Duty Alloys Ltd.

He asked me if I would join him, since I had helped in the development of Y-alloy forging and had in fact first suggested the type of salt bath for curing the forging blanks. The process maintained them at a constant and close temperature soak before forging. But I wished to remain with engines and, after my experience with the Stromboli and other motors, had become very interested in the inter-related problems of engines and their fuels. I had already been approached by Henry Tegner, Manager of the newly-formed Technical Sales Department of the Anglo-American Oil Company (now Esso), who wanted me to join him and deal with aero and automobile engine problems.

In my first year at Hooker's I had been in lodgings in Walthamstow with a very nice cockney family, going home at the weekends to my parents who were then living in Ealing. After our marriage in Whitley Bay, in September 1925, Con and I were offered part of my parents' house as a flat. This meant a very early start in the morning, 6 a.m. to get to Walthamstow by 8 a.m., and involved travelling the whole length of the Central London underground railway to Liverpool Street, from there by steam train to Walthamstow and then a tram to the works, all in all about four hours' travelling a day.

One day, after about a week of engine testing, and with my head and ears buzzing, I was on my way home on the Central London when, quite suddenly, I was assailed by a feeling of claustrophobia and felt that I must leave the train at the next station. But I reasoned with myself that if I succumbed to this and alighted I would never be able to travel on the underground again. So I forced myself to sit it out until Ealing where I left the train feeling thoroughly ill and in a perspiration. It took some weeks of sheer travel misery on the Central before the claustrophobic feeling went and I was fully in control of myself again. I give this merely as an illustration of how, after heavy pressure and difficult living conditions, one can come near to a nervous breakdown with little or no prior warning.

About 1926 Colonel Fell, who was in charge of engines at the Air Ministry, had the idea of an operating cycle to give high economy (i.e. low fuel consumption). The unit designed to test out the theory took the form of a horizontally opposed, two-stroke, double crankshaft, four-cylinder engine, having each pair of cylinders arranged one above the other with a common combustion chamber. The crankshafts were geared together at 2:1 ratio and their phasing was such that the two adjacent pistons arrived together at top dead centre. Then, after ignition, one went down at twice the speed of the other. The intention was to provide greater expansion of the charge and so produce more useful work from a given amount of fuel.

In practice, and partly because its design resulted in an extraordinarily low mechanical efficiency of around 60 per cent, Fell's hopes were not realized and, with the closure of Hooker's, the re-design and further development of this engine were stopped.

Some very good engineers were bred at Hooker's: Eric Moult, in the design office, who worked with Halford for many years and later with the Rolls-Royce Small Engine Division; Basil Stephenson, also in the design office, who became head of design at Vickers-Armstrong, Weybridge; Edward Chatterton, a designer at Hooker's, who went to Napier and became a very worthy successor to the late H. C. Tryon as a sound development engineer. After retirement he became a consulting engineer, but has since died.

Among the interesting jobs at Peter Hooker we built Parry Thomas's 1.5 litre engine for his low-built flatiron racing car which he drove in events at Brooklands. We also made the clutch and gearbox he had redesigned for his World Speed Record car, *Babs*, in which he was killed shortly afterwards at Pendine.

John Siddeley had offered me the position as chief engineer of the aero engine company while I was at Hooker's. The distinguished engineer from Farnborough, Major F. M. Green, had held the overall position as chief engineer of Armstrong Siddeley and Armstrong Whitworth but was to concentrate on the aircraft side, hence the offer to me. Siddeley paid his senior staff well but they were expected to do what he decided and this was often at variance with their own technical judgement, which meant that they had to accept direction or get out. When I went to see him after his letter of offer, he suggested that I looked around the engine test and experimental section. This I did and in the late afternoon went back to his office, mentioning that I thought about £200,000 should be spent on new and improved test facilities – knowing about this from past visits. Siddeley answered, 'We will talk about that when you join us'. I decided, however, that I would not make the change and was later to be glad that I refused since I would probably not have had the opportunities that came my way later.

9 *In Oil*

I joined the Technical Sales Department of the Anglo-American Oil Company almost immediately after Hooker's folded and there met Sam Heron who, to me, was one of the most capable internal combustion engine specialists, combining designing ability with a considerable knowledge of the chemistry of fuels – unusual in a mechanical engineer. He had been asked by Henry Tegner to see me before I joined and confirm that I had the experience and knowledge they wanted.

Sam, now dead, could be very tough with people. He was fiery tempered and irascible and certainly didn't suffer fools gladly; not even some friends who, at one time or another, came into disagreement with him. But he was exceedingly kind at heart when one pierced the armadillo-like shell, and would go to great lengths to give help and advice on all manner of engineering problems, though one never took anything for granted with Sam however long you might have known him. It was he who invented and developed the salt-cooled and, later, the sodium-cooled valve.

Sam Heron was a North East Coast man and his father had been a Shakespearian actor of some standing. A good deal of Sam's irascibility could be put down to the hip joint disease he suffered from youth and which kept him on his back for a number of years in early life. He was one of the team of engineers at the Royal Aircraft Factory (Farnborough) in the 1914–1918 War, working with Dr Gibson and H. P. Boot, under Major F. M. Green, who eventually advanced the design of the air-cooled engine. He went with Green to Siddeley's with the RAF8 (Jaguar) when that engine, or rather its design, was handed over by the Royal Aircraft Factory. But Sam did not stay long at Coventry since neither he nor John Siddeley could see eye to eye on how things should be done when the Jaguar was being developed, finally coming to complete disagreement when Siddeley wanted Sam to change from a two-valve to a three-valve head, presumably because the water-cooled 'Puma' they were building was designed with three valves – two exhausts and one inlet.

After leaving Armstrong Siddeley he went to the U.S.A. and was accepted, a Britisher, by the U.S. Army Air Corps (now U.S.A.F.) as a civilian research engineer on aero engines in the Engineering (Power Plant) Division at McCook Field, which subsequently moved to Wright Field. At first under Major Hallett and then E. T. Jones and Captain Tillinghast, followed by Lieut. (later Colonel) Eddie Page (with Major C. W. Howard whose efforts had made funds available for fuel research) Sam worked on the hot end of engines, mostly air-cooled, single-cylinder units; becoming interested in, and concentrating upon, the development of better performance from improved fuels. This proved to be his métier and indeed his life's work and interest. He went to Wright Aeronautical at one period and redesigned the cylinder of the famous J5. Sam personally inspected each and every component of the engine for Lindbergh's Atlantic crossing machine, the Spirit of St Louis.

In the period 1926–28, after the discovery of tetraethyl lead as an anti-knock by Midgley of Delco (a division of General Motors), the Anglo-American Oil Company in the U.K., who had until then merely offered automobile fuel (Pratt's spirit) on specific gravity and distillation range, made an agreement with the newly-formed Ethyl Gasoline Corporation to market a premium grade leaded petrol, then branded as Pratts ETHYL. The Anglo-American were to have exclusive rights for its sale on the British market for one year.

Heron was invited over to help educate us on this fuel anti-knock, with special regard to the engine problems likely to be encountered in service. He was on leave of absence from the Air Corps for some months and his advice proved invaluable. He was most interested when I told him of my own experience at Hooker's with tetraethyl lead. History would be incomplete if the uninformed newspaper campaign against leaded fuel were not mentioned, with the subsequent Government investigation into the alleged health hazards. The (late) Dr Graham Edgar, the distinguished American fuel chemist of Ethyl Gasoline and inventor of the octane number scale, with Dr Robert Kehoe, the medical adviser, both came over and gave evidence before the Government Committee which eventually cleared the use of tetraethyl lead in petrol.

Joining the Anglo-American Oil Company I learned something about the chemistry of hydrocarbon fuels but after a few months in London I was sent to Leeds to open a technical office there in order to deal with any possible queries and problems attending the marketing of Ethyl petrol in the North of England and Scotland. From Leeds, I covered the whole of the North of England, Scotland and, on occasions, Ireland.

After a year in Leeds, I was brought back to the London office and eventually became Assistant Manager of the Technical Sales

Department. My replacement in Leeds was a bright young physicist, Hugh Tett (now Sir Hugh Tett) who, after World War II, became Chairman of Esso in the U.K.

In 1930, I decided to qualify for my pilot's licence and went to my friend Duncan Davis who had then taken over from Colonel Henderson the flying school at Brooklands. Duncan, who has now passed on, was one of the brighter of Smith-Barry's instructors and, profiting from my previous flying with that great founder of modern pilot training, I was able to go solo in about two hours and obtained my Royal Aero Club certificate together with the Air Ministry A Licence after a total of some six hours' flying. However, forewarned by what I had seen of flying over the years, I continued advanced training with Duncan and, every so often, had a refresher course with him, Ted Jones or George Lowdell, according to their availability.

The Brooklands School of Flying was undoubtedly the best of the many flying schools in the U.K. of the period and, most important, it had an excellent ground staff who kept all the machines in first class condition. Duncan had given up Henderson's Renault engined Avro 504, in favour of Moths and Tiger Moths; only the latter were then used at Brooklands.

I had been with Anglo-American for less than two years when I was approached by Ray Bevan to join him. He had left the Technical Sales Department of Anglo-American to become manager of The Ethyl Gasoline Corporation (export division), set up to cover territories outside America where it was intended to introduce leaded fuel. Ray (T. R. A. Bevan), a Welshman, was up at Cambridge and had left there with a B.A. (Engineering) to go to the Anglo., where he served in a number of departments including Sales. Having an engineering degree, however, he was early co-opted into the Technical Sales Department.

Bevan asked me if I would look after the engineering side of things, since he was largely occupied on administrative and commercial matters. This was exactly the type of work I thought I would like, particularly to be in at the start of a new development. Rarely does such an opportunity occur. Further, I have always preferred to work in a small organization where ideas and decisions can be more quickly developed and agreed, or not, as the case may be.

10   *Tetraethyl Lead*

I joined Ray Bevan and the Ethyl Gasoline Corporation in November 1930 and from then onwards was able to use my engineering background to gain more experience in a new field; with every encouragement and no hindrance on Ray's part. In 1931 we became the Ethyl Export Corporation. We were concerned with the application of fuel to aero and automobile engines and this work embraced the principal aero engine and automobile manufacturers in the U.K. and Europe, including the military air forces and commercial airlines.

As a side issue, while with Anglo-American, I had taken an interest in car racing fuels, since this activity had an important bearing on keeping the name 'Pratts' and 'Ethyl' in advertising and sales. A number of the racing fuels used by Anglo for some years were blends of basic mixes I had prepared and tried at Brooklands and elsewhere in road races, for both cars and motor-cycles.

The events leading to the discovery of tetraethyl lead as a fuel anti-knock are interesting and one has to go back to 1917 and the particular circumstances which started the American investigation in the anti-knock field.

In that year, the shortage of petrol (gasoline) because of war demands caused a change to kerosine (paraffin) as a fuel for house lighting sets in the U.S. However, there soon came widespread complaints of engine knock or detonation and overheating. The Delco Corporation was the major supplier of house lighting sets, so they referred their problems to the General Motors Research Corporation whose head was Charles (Boss) Kettering. Kettering, who had been familiar with the knock or detonation problem as early as 1911 and had a theory that associated the phenomenon with the fuel, handed the Delco problem to two mechanical engineers on his staff, Thomas Midgley Jnr., and T. A. (Tab) Boyd, with the following instruction: 'Find the cause; cure it'. Midgley, Boyd and their associates were soon able to satisfy themselves that Kettering was right in associating detonation with a fuel characteristic.

The first theory explaining the mechanism of detonation of

fuel came from H. R. Ricardo (the late Sir Harry Ricardo) at the end of the First World War. Ricardo had worked with the late Professor Hopkinson and following the latter's experiments with cylinder indicators had deduced that detonation was a fuel phenomenon. Ricardo came to the conclusion that the conditions causing detonation were due to radiation from the flame front, initiated at the sparking plug as it advanced across the combustion chamber. This heated the yet unburned portion of the charge which was also supercompressed by the burning and expanding gas behind the flame, to a point where the end (unburned) charge spontaneously detonated before the flame reached it. The burning charge can be likened to the relatively slow burning of cordite while detonation is akin to that of fulminate of mercury. When the British workers, Ricardo, Tizard and Pye started their investigation of fuel behaviour, accepting Ricardo's theory, it was known and confirmed by them and the American workers that fuels of various hydrocarbon or chemical structure behaved differently under the same engine cylinder operating conditions. This was, of course, known some years before either country got to grips with basic problems.

To test the radiation theory, the GM team added aniline dye to the fuel to see if it would mask or cut down the radiation. Iodine was also tried. Some anti-knock effect was actually observed, not, however, as a result of a reduction in radiation but rather due to an inherent chemical reaction of an anti-detonating nature.

Dr Wilson of the Standard Oil Company of Indiana, later to be its President and then Chairman, made what was probably the most valuable suggestion to Midgley, which was that a study of the Periodic Law by Mendeleyev of the atomic weights of the elements might be worthwhile.[1] Pursuing this study Midgley found that the elements having the higher atomic numbers, or weights, seemed on test to have increasing anti-knock effectiveness. This led, almost naturally, to metallic lead as an obvious choice for further investigation, but the difficulty was then to find some method of converting the lead into a form which could be used in gasoline.

From this came the organo-metallic compound, tetraethyl lead, a heavy, oily, colourless liquid easy to mix with gasoline. In that form, before combustion, it is a highly dangerous and toxic material, but this characteristic was not at that time given serious consideration. In fact, tetraethyl had until then been virtually a chemical curiosity.

Laboratory quantities of tetraethyl lead were made and engine tests confirmed its remarkable anti-knock effectiveness. But the lead

---

[1] Dimitry Ivanovitch Mendeleyev (1834–1907 produced his Periodic Law (Table) of the atomic weights of the elements in 1869.

oxide, formed during combustion, tended to foul sparking plugs and cause or accelerate exhaust valve burning. Midgley's agile mind considered that the addition of a halogen bearer might reduce these undesirable deposits by combining with the lead upon combustion to form, say, lead bromide, which was relatively volatile and better evacuated from the cylinder during the exhaust stroke. Experimenting with brominated compounds, which were added to tetraethyl lead before blending in the fuel, he eventually came upon ethylene dibromide and, later, ethylene dichloride. The latter was somewhat cheaper to produce and was added to motor gasoline while the former, being more effective as a scavenger, went into aviation fuel.

In view of the early success of Ethyl Fluid and the need for properly controlled manufacture in quantity, a company was formed and named the Ethyl Gasoline Corporation (later, Ethyl Corporation), equal shareholding being held by General Motors and the Standard Oil Company (New Jersey). Ethyl Gasoline was formed to retain technical control and to market tetraethyl lead in the form of Ethyl Fluid. The du Pont chemical concern, at that time a major General Motors' shareholder, was responsible for the manufacture of lead for Ethyl Gasoline. Later, Ethyl Gasoline manufactured their own lead. But after the Second World War, when the manufacturing patents had run out, a number of other companies took up the manufacture of TEL.

The success of tetraethyl lead over the years, apart from its effectiveness, has been in large measure due to the strict medical control exercised during its manufacture, blending and concentration in gasoline.

At about the same time as Midgley and his associates were investigating detonation and its control, German chemists and engineers were engaged in similar work. The Badische Anilin & Soda-Fabrik (later allied to the I.G.Farbenindustrie A.G.) were the principals involved, the early work being done by Dr Mueller-Conradi, a chemist, and Professor Wilke, an engineer. They produced tetraethyl lead in laboratory quantities and approached Daimler-Benz, who co-operated in a series of engine tests using one of their 4 cylinder, 2.5 litre, engines. But during their researches, Mueller-Conradi came upon iron carbonyl, also a very effective anti-knock, which Midgley had already tested. Following this discovery, BASF in June 1924, established an engine laboratory at Oppau under Professor Wilke's management. In June of the following year, Richard Morgenthaler, an engineer in Daimler-Benz, transferred to BASF and worked with Professor Wilke.

Since patents covering the use of tetraethyl lead as an anti-knock had already been granted to General Motors Corporation, Wilke and his associates concentrated upon the development and use of iron carbonyl. Owing to Midgley's work, the patent position regarding the use of iron carbonyl was obscure. Subsequent discussions between General Motors and I.G.Farbenindustrie resulted in a mutual agreement to allow the German company exclusive rights to iron carbonyl as an anti-knock compound within their own company, and motor gasoline containing iron carbonyl was put on sale in Germany in 1926.

Although it had good anti-knock properties, the use of iron carbonyl resulted in increased engine wear (rusting of cylinder bores and piston rings) and shorting of the sparking plugs. Ethyl Gasoline, themselves, had put considerable effort and funds into trying to overcome these difficulties since iron carbonyl was cheaper to manufacture and less toxic than tetraethyl lead. But they abandoned it. All this was to be confirmed later by German experience when gasoline containing iron carbonyl was sold to the general public and then had to be withdrawn.

The dropping of iron carbonyl did not, however, immediately react to the benefit and more general acceptance of tetraethyl lead in motor fuel in Germany, since neither of these organo-metallic compounds could freely be used by the oil companies because of a Government monopoly on benzole and alcohol production. These ingredients, if in surplus, had to be used in motor gasoline, though it was difficult to control the anti-knock value of the marketed gasoline to within reasonably close limits owing to the varying amounts of benzole and alcohol available at any one time. This lack of consistent quality was largely responsible for renewed interest in tetraethyl lead in 1934, in Hitler's Germany, since its advantages were apparent, compared with benzole and alcohol, for aviation fuel.

In 1934, the Royal Aeronautical Society invited me to give a Paper on ETHYL for which I was later awarded the Taylor Gold Medal.

Soon after joining Ethyl Export I went to the States, to visit the Ethyl Gasoline Laboratories in Detroit and to learn what I could of tetraethyl lead and fuel knock testing. There was also a laboratory concerned with fuel chemistry at Yonkers, just outside New York City. Here, Drs Edgar and Calingaert worked. In Detroit I visited General Motors Research Laboratories and met Boss Kettering.

It was the beginning of the 'thirties, America was still dry, and my new friends and colleagues of Ethyl Gasoline took me to a speakeasy somewhere on 42nd Street near to their new offices in the Chrysler Building, where they had just moved from downtown (26) Broadway. After drinking and talking for some time, there was a sound of two shots outside and the speakeasy staff hurriedly collected all the drinks and locked the doors. We were shut in for about an hour, listening to the wail of police car sirens. When there had been comparative quietness for a period, they opened up for business again. It transpired that a policeman had shot it out with a suspected gangster who had taken shelter in the speakeasy next door.

While learning about the chemistry of tetraethyl lead and fuels in the Detroit Laboratory, we had occasion to visit Chicago; I was driven there by Julian Frey accompanied by his new and attractive wife, Muriel. We were met in Chicago by the local Ethyl representative, who thought we ought to see the night life. Our entertainment ended in a night club in the Cicero district, which was the area or domain of Al Capone. The club was a terrible place to take a young girl and I was most embarrassed for Muriel. There were near-naked girls dancing on the tables and some awfully low turns, although today they would probably be regarded as pretty old hat.'

When it was time to pay and go, for which I was truly thankful, the bill presented was about three or four times more than we dreamed it could be, even allowing for the type of joint we were in. Our Chicago man started to remonstrate, but when we saw a number of very tough chuckers out advancing towards our table I suggested we pay up and get out since that would be cheaper than the subsequent hospital fees if we mixed it. Never was I more thankful to leave; though having been in almost equally rough places elsewhere in the world, I felt quite defenceless with a girl to look after into the bargain.

There was no doubting the difficulties with exhaust valves at the start of marketing leaded fuel, both in the States and in the U.K. But whereas in 1926 there was greater and more general distribution within the U.S., it was little used in the U.K. from its introduction in 1928 up to the beginning of the Second World War. The Anglo-American Oil Company enjoyed a year's monopoly and when other concerns, notably B.P., came in, only a few per cent of the total fuel sold in the U.K. was premium Ethyl. It was difficult to promote and increase Ethyl sales with the availability of alternative grades and brands of unleaded petrol, particularly when competitor oil companies were over-emphasizing engine difficulties with leaded fuel.

Ray Bevan and I were of course very much involved with our customer companies, but were determined that we would always be technically honest and indulge in no sales evasions. So, from the start, we treated all problems strictly from an engineering standpoint and, I hope, objectively. We helped our customers, the oil companies, to identify the various types of valve failure, indicating how trouble could sometimes be aggravated by poor engine maintenance and incipient valve sticking and burning before even leaded fuel had been used.

At the same time, visits were paid to the car manufacturers and valve makers in an endeavour to get improved valve materials accepted with better valve and port design. This, of course, was not easy in those earlier times of quantity production, where any alterations affecting cost and production were naturally most strongly resisted. In those days, control of materials was not always very rigid and it was quite usual for the buyer to order valves to a pretty wide specification and accept those most competitive in price. I used to say at the time that exhaust valves were often made from bumped-up iron railings, and some behaved little better in the engine.

Few motorcar manufacturers in the 'twenties and 'thirties had strong or comprehensive design and engineering departments, policy being dictated by a generally non-engineering Board or the Sales Director. Well-engineered vehicles usually only came from engineer controlled firms like Rolls-Royce, Lanchester, Riley, Vauxhall and Sunbeam. But few, if any, of the engineering staff of other companies reached Board level. Things have changed for the better in recent years, especially since the end of the Second World War.

Except, possibly, for the older specialized and high class car firms in the U.K. and Europe, such as those already mentioned and Hispano-Suiza, Delahaye, Talbot, Hotchkiss, Fiat and Mercedes-Benz, not forgetting Jaguar and Aston Martin, quantity production cars were not conspicuous by good engineering detail. General Motors and Chrysler in the U.S. were probably the first to lead the way in better car engineering in the quantity production field.

Without doubt the side valve lasted too long. In fact, side-valve engines were still being manufactured by many firms up to the Second World War. Whatever the engineering effort put in to improve their operating conditions and life, they were always inferior to the overhead type; although these did not always contribute as much as they could have done to valve life and performance owing to poor design. Improved exhaust valve materials were not readily available except, possibly, in the more costly aero-engine valve steels of the austenitic variety. Later, however, for automobiles, came the Silchrome steels and in recent years the XB and the 21-4NS varieties.

So far as the sale of leaded fuel outside the States was concerned, the years before the Second World War were mostly spent in the education of the British and European automobile industries and oil companies on the advantages of leaded fuel as the low cost refinery anti-knock additive. But a lot of useful work was done with the aero-engine manufacturers in developing engines to take advantage of such fuel. I concentrated my efforts in this direction since I was convinced that the experience gained in military and commercial aviation would stand us in good stead for the general use of leaded fuel in automobiles later.

We of Ethyl Export were, in a sense, consultants and laid ourselves out to help the aero-engine and automobile industries in coping with any problems they had, in particular those to do with the 'hot' end of the engine. Often, though, my work was also concerned with the solution of various other engine operating problems. By this means we became accepted as engineering advisers, and this policy helped us considerably when leaded fuel became more generally accepted since one was already *persona grata* with the Industry.

In this connection, I also had very good contacts with the American aero-engine and automobile industries and Sam Heron had by the middle 'thirties left Wright Field to become the aviation research engineer for Ethyl Gasoline in Detroit, which was a great help to us.

The emphasis on the importance of technical honesty was only commonsense, for if one hoped to be accepted internationally it would have been stupid to descend to flanneling. This is not to say that one had contempt for the hardworked salesmen of the oil companies, but their enthusiasm sometimes overshadowed their judgement. It was often difficult when visiting a complainant with a sales representative to give a balanced view without letting down the salesman. However, by our constant contacts with the oil companies' technical and sales staffs, an atmosphere of mutual trust was built up and complaints were dealt with in a straightforward manner.

11 Engine Progress

The rate of progress of the internal combustion engine in the latter part of the last century and this century was broadly determined by restrictive legislation, particularly in the U.K. It started with the man with the red flag walking in front of the car. Then later it's engine was controlled by the R.A.C. horsepower rating formula of 1906.[1] This tended to limit cylinder diameter and, as engine knowledge advanced, the valving. But today, with the R.A.C. formula gone in favour of flat rate taxation, we see square and oversquare engines, with equal or larger bore to stroke ratios and V-type engines of 6, 8 and 12 cylinders for compactness. The evolution of the automobile, from a powered dogcart and brougham to an integrated vehicle started early in the present century.

The aero-engine developed rapidly after the First World War when, at the War's end, two distinct types emerged which were to dominate the scene until the piston engine was made obsolete in aviation by the gas turbine for all but aircraft of the lowest power. The two basic categories were the V-type, 12-cylinder, liquid-cooled engine and the single or double-row, air-cooled radial. Examples of the two-row engine were designed during World War I but did not see war service. There was also the Napier 'Lion', an arrangement of three banks of four cylinders in $\downarrow$ (broad arrow) form, a successful liquid-cooled engine, designed in that War, but first used in the immediate post-war era. A similar type of engine was made by Lorraine and Farman in France.

Of the later radials, there were the single-row ABC (but the less said about that engine the better), the Jaguar and the Jupiter.

The 14-cylinder, two-row, air-cooled Jaguar, built by Armstrong Siddeley, competed with the Cosmos Jupiter, a 9-cylinder, single-row, air-cooled radial, designed by Roy Fedden and his team at the

---

[1] RAC rating was $0.4 \times D^2 \times N$, where 0.4 was a co-efficient, D was the cylinder diameter in inches and N was the number of cylinders.

Cosmos Engineering Company. It was formed by Fedden after leaving Brazil Straker of Fishponds, Bristol, when that firm gave up aero engine manufacture. Fedden was chief engineer of Straker's and responsible for their building water-cooled Rolls-Royce engines during the war.

When Fedden branched out to design his own aero engines he was committed to the air-cooled type, since Rolls-Royce had apparently imposed conditions upon him, when he took on the wartime manufacture of their engines, that he would not compete in the water-cooled field.

In many ways, the Jaguar was of more advanced design than the Jupiter, but since John Siddeley (Lord Kenilworth) never appreciated the need for intensive development, the Jupiter, by sheer development, gained in performance and acceptance. The Jaguar, however, was the forerunner of the classic two-valve, air-cooled radial to be so successfully developed later by Wright Aeronautical and Pratt & Whitney in the U.S.

The Jaguar engine had, in fact, the first mechanically driven supercharger, but Siddeley would not use it as such and had the gear train removed. With the impeller driven directly from the crankshaft it was called a mixing fan. In the event it wasn't all that good as a mixer to give a homogeneous charge. Again, when later the supercharger was accepted and generally used, Armstrong Siddeley engineers designed an advanced two-speed supercharger gear but this too was held back.

The testing and constant refinement of an aero engine is essential to its successful evolution and long life. Assuming an acceptable design, it is this intensive development that decides an engine's future and approval by the aircraft designer and manufacturer.

In those early post war years Fedden of Bristol and Hives of Rolls-Royce impressed me with the way their aero engines were evolved. And both, in their somewhat different ways, showed the aviation world what could be done with good leadership, first class teams and unremitting work.

12 Schneider Contests: Influence on Military Aviation

Probably the most important series of events in aviation between the wars occurred in 1927, 1929 and 1931 with the British Government's sponsorship of the Schneider Trophy Contests of those years. This resulted in three consecutive wins by Royal Air Force teams and the Trophy being won outright for Great Britain. These three contests were to become significant landmarks, where the importance of fuel in permitting improved engine performance was confirmed and subsequent engine and fuel development were to go hand-in-hand.

America can be thanked for a very sporting gesture in 1924 since, with no entries ready from the U.K. and Italy, they could by a flyover have had a second consecutive win, but they elected to cancel the Contest for that year.

My first active acquaintance with the Schneider Trophy Contests came in 1927 when I was with Peter Hooker. I was asked by H. C. Tryon, the experimental engineer of Napier, if I would come over to Acton and give them some advice on the high-compression racing Lion engines they were preparing for the Schneider Contest to be held at Venice that year.

They were using a light cut (volatile) gasoline with a high content of tetraethyl lead and were having plug trouble, aggravated by the high compression ratio of 10:1.

I was able to assist with advice on sparking plug modification, and also in resolving troubles due to the burning of pistons and exhaust valves. From then on Tryon and I became firm friends and of mutual help on engine development.

I went to Venice and saw Webster of the R.A.F. High Speed Flight team win, piloting the Supermarine S5 fitted with a geared Napier Lion. With others, including the Crown Prince of Italy, we were on the roof of the Excelsior Hotel when Guazzetti, the pilot of one of the Macchis, half-blinded by fuel from a fractured pipe, turned towards the hotel and we all ducked as he flew directly at us. Fortunately he leapfrogged the hotel to land in the lagoon on the other side.

At Venice I first saw Roy Fedden's *Mercury* engine. Although preceded by a motor of the same name, this Mercury started life as a racing engine and was fitted in the Short-Carter-Bristow *Crusader* float monoplane, the only private entrant in the Contest. This clean-lined machine was designed by George Carter, and the Mercury engine by Fedden's Bristol team. The engine, a supercharged nine-cylinder, air-cooled radial, had a magnesium crankcase for lightness and gave over 800 bhp. Most unfortunately, in rigging the aircraft upon assembly at Venice, the aileron controls were crossed, causing it to crash on its first test flight and seriously injuring Schofield, the pilot. This was a great pity since, though slower than the other British and Italian entrants, it had an interesting cowling arrangement, each cylinder being separately enclosed in a helmet type cowl with vertical cooling slots. When the wreckage of the machine was recovered the crankcase was fizzing like an Alka-Selzer, due to salt water action on the magnesium.

My first contact with Rolls-Royce was in 1927, where I met A. J. Rowledge who had designed the Napier Lion. He left Napier shortly after the First World War to join Rolls and assist in getting the Condor, an overgrown Eagle VIII, out of its various mechanical troubles. I also met Hives (the late Lord Hives) at that time, when he

The engine in the Short-Carter-Bristow Crusader Schneider Trophy racing seaplane of 1927 was developed to become the Bristol Mercury.

was manager of the experimental department of Rolls. Rowledge was, in my view, with Henry Royce, A. G. Elliott and 'Bunny' Butler of Bristol, one of the world's 'compleat' aero engine designers who could create a design and supervise its execution to the last detail.

For the next Schneider Contest held in 1929, Mitchell of Supermarine, who had been given an opportunity of designing a racing seaplane, decided that a larger and more powerful engine than the 'Lion' would be needed. Rolls were therefore invited by the Air Ministry to develop and build it. So it was rather remarkable that, when first approached, the Managing Director of Rolls, Basil Johnson (brother of Claude Johnson, the famous co-ordinator and first Managing Director of Rolls-Royce) wanted none of it and felt their place remained in the automobile field, despite the success of the firm's aero engines (Hawk, Falcon and Eagle) in the First World War. But Henry Royce and others decided otherwise and it was not long after this that Basil Johnson left and Sidgreaves took his place. A great opportunity, which could have been missed, was firmly grasped. It might be said, though it must remain conjecture, that had Rolls not developed engines for these two contests they would not have been in the commanding position they were to enjoy during the rearmament period and in the Second World War, with the Merlin in the Hurricane, Spitfire, Lancaster, Mosquito and Mustang.

The engine chosen by Rolls as the basis for conversion to racing rôle was the H, or *Buzzard*, a 12 cylinder 60° V, motor of 36.6 litres capacity developing about 825 bhp.

Known as the 'R' engine in racing form, it was strengthened in certain respects and fitted with a large double-sided centrifugal supercharger (to reduce diameter) which increased the sea-level power to 1,850 bhp with the engine operating on 100 per cent benzole. But, during test and endurance running, troubles began to manifest themselves at the hot end, with power fall off, exhaust valve distortion and burning and plug sooting on idling that sometimes prevented the engine being opened up to full throttle. Rowledge (Rg. as he was known by the Rolls coding of names) asked me to come and help. I had by then joined the Technical Sales Department of the Anglo-American Oil Company.

In 1929 Vic Halliwell was in charge of 'R' engine testing and development at Derby, assisted by an energetic Ray Dorey. In the following year (1930) Halliwell lost his life with Segrave while attempting to raise the World Water Speed Record in *Miss England* on Lake Windermere.

I am indebted to R. R. Rubbra (Rbr.) of Rolls-Royce for an interesting little technical story on the R engine, concerning its connecting rod design. The Buzzard engine, from which it was

Rolls-Royce Buzzard V 12 engine of 36.6 litres cylinder capacity developing 825 bhp.
It was converted to become the famous 'R' engine of Schneider Trophy fame. By
permission of Rolls-Royce Ltd.

Rolls-Royce 'R' engine. By permission of Rolls-Royce Ltd.

derived, had the standard Rolls fork and blade rod arrangement and this was also used in the first (1929) 'R' engine, but proved weak because the arch of the fork was clear of the big end bearing shell. The fork was highly stressed and there were some fractures during the 'R' engine tests. In the engines for the 1931 Contest a master and articulated rod design was successfully used, though there was the disadvantage of unequal cylinder block dimensions with a compromise on ignition timing. In fact, the stroke on one side was nearly $\frac{1}{4}$ in longer than the other and, initially, the upper piston ring overrode the top of the bore! The Merlin reverted to the fork and blade rod arrangement, but with the fork bedded to the bearing shell.

With only little more than a month to go before the Schneider Contest, I did not have time to do very much; however, by the simple expedient of diluting the benzole with a 'light cut' Rumanian leaded gasoline, it was possible to get the engine through its tests satisfactorily and race-ready – the contest being won by Waghorn piloting the Supermarine S6A with the 'R' engine.

The night before the contest provided almost a greater degree of excitement for those directly concerned than the actual affair the next day. A pre-race inspection of the newly-installed race engines revealed some aluminium on the plugs removed from a cylinder in Waghorn's machine. This indicated that the particular piston was probably picking up. According to the Schneider Trophy Contest rules, once the race engines were finally installed they could not be removed; but the Royal Aero Club Contest Committee members at Calshot agreed that components could be removed and replaced *in situ* in the aircraft.

There were some hundreds of personnel from the Rolls Special and Experimental Departments, who had built and tested the engines, in Southampton that evening. They had come to view the event, but were spread all over the town in various hotels and pubs. It took some time to locate and round up the specialists among these skilled people who could remove the cylinder block concerned. Eventually, a team was gathered together under Lovesey, the senior experimental engineer. They worked through the night, removed the cylinder block, changed the offending piston, replaced the block and had the engine running at 8 a.m.

Waghorn, who won the Contest, was not told until afterwards about the night's happenings.

Since the RAF High Speed Flight team had won two consecutive Contests, and three successive wins by any one country secured the Trophy for permanent retention, it was obvious that the next Contest was of supreme importance to us. This was arranged for 1931, again at Calshot, with the Italians, the principal challengers, known to be working on some quite advanced and even revolutionary seaplanes – one, the Piaggio P7, using its fuselage to float directly upon the water, like a boat, with hydrofoils and a marine type propeller to raise it into the take-off attitude before the main (air) propeller was clutched in. Another was the tandem engine and propeller Savoia Marchetti S65, with an engine at each end of a short fuselage or nacelle and an outrigger tail.

Imagine therefore the alarm and despondency in the aircraft industry and sporting aviation circles when the British Government of the day announced that, because of the country's economic state (the Depression was very much with us), the contest could not be officially supported. And the situation was not helped by Trenchard's apparent indifference, even prejudice, towards the High Speed Flight. However, Lady Houston, the widow of a shipping magnate, shamed the Government into revising this shortsighted decision by offering to contribute £100,000 towards the expenses.

All this 'kerfuffle' meant that there were only a few months left for Rolls to increase engine power to the 2,350 bhp needed by Mitchell, the designer of the S6B, if he was to attain the performance thought necessary to beat the best of the Italian machines. In the event, the Italians did not turn up and Boothman, piloting the S6B, won the Trophy outright for Great Britain at a speed of just over 340 mph.

There were amusing, and sometimes not so amusing, highlights at the 1929 and 1931 Contests when Rolls came into the picture. For instance, it was an experience to drive with Hives from Derby to Calshot on one of the many trips made between these two places during the experimental and practice flying phases. Through his previous background, Hives was a fast driver and his well-known sixth sense seemed to work when he drove a car. I also drove fast in those days, but sitting beside him while he took the best line on a corner on a country road was more than somewhat nerve-racking. When he did this rarely was any oncoming traffic met. On another corner he would refrain, becoming relatively cautious, and sure enough a car would be coming in the other direction.

Hs. (Hives) liked snooker after dinner, and when Rolls acquired their guesthouse at Duffield Bank and he came to dinner with us, his guests, we would always repair to the billiard room afterwards. He was not all that good but many of us were much worse and I was often paired with him. But if we were beaten in the first game he dragged me along with him by sheer force of personality to win the next, and often the rubber. He just would not accept defeat in anything, once he had decided it was worthwhile to fight.

When our participation in the 1931 Contest had been agreed, I was due to sail for the States and so I left Ray Dorey (then responsible for 'R' engine development) with some suggested fuel mixes for the initial test running. Hives gave me a detailed drawing of the 'R' engine exhaust valve and asked me to consult Sam Heron, get two engine sets of sodium-cooled valves and, if possible, bring them back with me. Sam and I went to see Robert Jardine of Wilcox-Rich (who worked Sam's original patents for the sodium valve) and he promised to produce the valves while I completed the rest of my American visit programme. These valves proved most successful in the 'R' engine tests. Under Schneider rules it was necessary for the complete engine to be manufactured in the country of origin, so Rolls obtained a sub-licence from Bristol (who had originally taken the British licence for the sodium-cooled valve) and manufactured the valves themselves. They were stem-cooled, not the hollow-head variety.

Rolls-Royce sent, and collected, their practice engines to and from Calshot by road, and this was done by using a Phantom I car chassis fitted with a cradle for the engine. There were many records claimed for the times between Derby and Calshot which now escape me, but I do know that the police stopped them once after they had toured through a village at some 80 mph.

The shortage of time to prepare for the 1931 Contest was a cause for mounting tension, though this was not particularly apparent at Derby. But barely a month before the event the 'R' engine had not

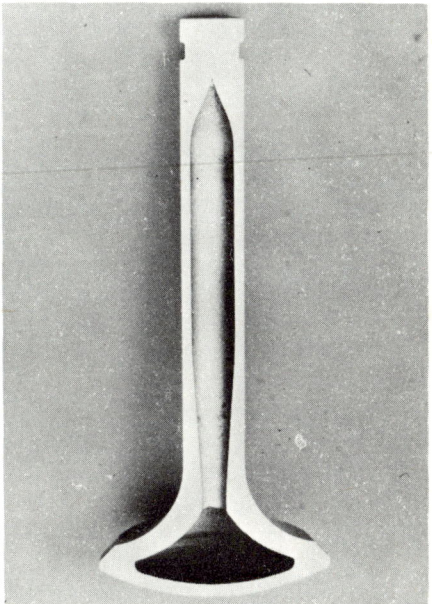

Hollow-head, sodium filled valve (to 60% of internal volume), used in large air-cooled radial engines. The Rolls-Royce 'R' engine valves were stem filled only, to 60% of internal volume.

completed the official one hour endurance run at full throttle agreed between Rolls and the Air Ministry. The first attempt was a few days before the August Bank Holiday, but after 34 minutes the crankshaft broke! Since the crankshaft is a somewhat important component of an engine, this provided food for thought. However, it was discovered that the particular shaft had already exceeded the arbitrary life of, I believe, seven hours, so an engine was built up with a new shaft and another start made on the Friday of Bank Holiday eve. To the consternation of all concerned, this shaft broke after 58 minutes – only two minutes short of the hour! What to do?

Rowledge, in his quiet way, decided that the only alternative open to him in the time was to leave more material in the shaft by way of smaller bores in the pins and journals. Coverley, head of the Special Department responsible for building the engines, motored that evening to Sheffield, visited the steel company concerned and got hold of the manager who gathered a forging team together. By working all through the holiday period Rolls got new forgings, the first of which was rushed to Derby for machining that weekend. Within a week an engine ran the hour at full throttle and passed the test.

It was usual after each contest, while the High Speed Flight team and equipment were in being, to make an attempt on the World Absolute Air Speed Record. This had been done in 1929 by the S6A, piloted by Orlebar (Commanding Officer of the High Speed

The Rolls-Royce 'R' engine test team at Derby, 1931. Left to right *Stewart Tresilian, Bob Young, Stan Orme, Frank Nicholls, Bill Thacker, Charlie Conway George Parkin, Ray Dorey and the Author.* By permission of Rolls-Royce Ltd.

Flight), at a speed of 357.7 mph. In 1931, Stainforth had, after Boothman's flyover, raised the Speed Record to 379 mph, but Hives was not yet content and with his usual foresight said that the next mark that mattered was 400 mph and not 399.9 mph. The Air Ministry, completely lacking in imagination, wished the High Speed Flight to vacate Calshot and hand the base back to the flying boat squadron that had vacated it. This caused a furore, but the decision was reversed after Sir Henry Royce interceded on behalf of the High Speed Flight.

I went back to Derby and Hives asked me what could be done to provide more power by altering the fuel, since there was no time to modify the engine. I thought another 200-250 bhp was possible if the resulting high fuel consumption was not excessive for the tankage with the six runs up and down the speed course – also, perhaps, the supercharger gear ratio might be raised.

Dorey and I mixed up a high alcohol content fuel 'cocktail' and this, with a higher blower ratio, gave 2,800 bhp. But things started to happen to the engine and some cylinder holding-down bolts failed, so we reverted to the original ratio and the engine gave about 250 bhp increase on the fuel alone.

The final 'sprint' fuel was a blend of 60 per cent methanol, 30 per cent benzole and 10 per cent acetone, plus lead. Therefore, the consumption was going to be much higher than that with the contest fuel. On the first bench test with the sprint fuel, we noted a 'swing' in rpm with a tendency to 'cough' back into the super-charger and it was found that the correct jet size demanded a fuel flow larger than the fuel pump was capable of giving! The discovery was made one evening and the remedy was to step up the gear ratio of the fuel pump drive. This involved a chase around Derby to find the man responsible for the gear cutting section. He was finally located and prevailed upon to set up the machines to cut the gears. It was all done very quickly and, with the engine running again, the fuel tests were completed in a day or so.

As soon as the Air Ministry agreed the speed record attempt the engine was rushed to Calshot but difficulties arose with tank sealing when it was installed, since the fuel I had concocted proved a very good solvent of paint and the sealing compounds. Some time had therefore to be spent in getting suitable substitutes for the original preparations. Treated cork seals were eventually used and also different sealing compounds for the tank jointing. When all was apparently ready there was further difficulty with the carburation of the fuel: a good deal of throttle had to be used to keep the engine running and avoid the plugs damping out. Finally, a drill was evolved whereby the machine was put on its launching lighter and towed to the take-off point. The engine was warmed up on the lighter at fairly large throttle opening, then shut down and the induction system drained. Coverings were immediately placed on the wing and float surfaces, which formed the radiators, and on the tail fin containing the engine oil.

When all was ready for the attempt, the pilot, Stainforth, was settled in the cockpit, the coverings removed and the engine started. He then opened the throttle and fairly shot off the lighter for the take-off. On reaching the end of each run on the speed course, he throttled back somewhat on the turn towards the next run and wet fuel vapour was seen coming from the exhausts! But the magic figure of 400 mph was surpassed by 7.5 mph (407.5).

Neither before the 1929 and 1931 Contests nor since have I been involved in an operation which sparked such individual and collective energy and enthusiasm.[1]

Rolls-Royce gave a large party after the final Schneider Contest, and I motored the pilots to Derby. I was driving the latest Buick and Stainforth particularly wished to try it. Notwithstanding warnings from the others I let him, but after a few minutes we had had enough so I reassumed control. It has often surprised me that a number of first-class pilots were such bad car drivers. Stainforth, who had great piloting ability and judgement, had none as a car driver, and he was always in trouble with his little 7 hp Austin at Calshot. What he would have been like in present-day traffic conditions I shudder to think.

My work on the Schneider fuel with Rolls started a long association with this famous firm and I am lucky to count many friends made there over the years.

The 'R' engine development showed what could be done with high anti-knock fuel and supercharging to get more out of a given cylinder capacity, and indeed to keep the engine from growing unduly in size. Until that time there had been little actual supercharging, the supercharger or blower being used to restore atmospheric or sea-level power at altitude. Subsequently, with the advent of 87 octane and, later, 100 octane fuel, it was possible to increase considerably the power of an engine and reproduce this at high altitudes. In the Second World War, the power could be maintained at still greater altitudes by means of two-speed and, also, two-stage supercharging, together with after-cooling of the charge between blower and engine cylinders.

Actually, the fuel used in the Schneider 'Lion' engine of 1927 was more practical for normal and general service use than the fuels I prepared for the 1929 and 1931 'R' engines; and it was the lack of any charge cooling that caused me to add alcohol in order that its higher latent heat (compared with that of gasoline) would reduce the charge temperature in the blower and get greater charge weight

'*R' engine on the test bed at Derby with the ram air intake fitted. Ram air was supplied by a blower driven by a R-R Kestrel engine.* By permission of Rolls-Royce Ltd.

into the cylinders. This worked well at little expense in fuel consumption so far as the contest fuel was concerned, but the sprint fuel for the speed record would have been quite impracticable for normal use. It was indeed a 'cocktail'. It showed, however, that a temperature sensitive fuel could be used with advantage in a cool engine. But such fuel would probably have been of little use in an air-cooled motor, and might even have pre-ignited.

It was from my association with engine development in these last Schneider Trophy Contests that I was persuaded to write my first journalistic article for *The Aeroplane*. I had met its famous editor, C. G. Grey, during the 1914–1918 War, through a mutual friend of my father's, and that started my long friendship with this remarkable man.

C. G., as he was universally called, was a forceful personality and his editorials were worth reading, though one did not always agree with all his views. But he was vital and always interesting.

C. G. asked me if I would write an article for *The Aeroplane* on the development of the 'R' engine. I said I would but he could reject it if he didn't like it. The article, *The Evolution of a Schneider Engine*, happily a great success, was published in the issue of the 7th October, 1931. Grey was delighted.

I took no fee, for two reasons: first, I wanted to do it for all those with whom I had collaborated and, second, C. G. had the devastating habit of altering a word or a line, or even a paragraph, in a contribution, which often gave it quite a different angle or meaning. I was determined that this was to be my article and mine alone, so I got an undertaking from him – no fee and no alterations.

Rolls-Royce's publicity department had some five thousand reprints made and distributed.

[1] *The Schneider Trophy Races* by Ralph Barker, Published by Chatto and Windus, 1971; and *The Schneider Trophy* by David Monday, published by Robert Hale, London.

13 *Italian Interlude*

After the flyover by Boothman, giving us permanent possession of the Schneider Trophy, the Italians decided to try to take the Absolute Air Speed Record from us; and I was, in 1932, asked for by General Crocco, the doyen of Italian aeronautical engineers, to come to Italy and help in the engine development. This approach was made through the oil company supplying the fuel, by their aviation representative, Colonel Guerritore, a retired artillery officer.

Apparently, Mussolini had decreed, like the Almighty, that in the record attempt, there would be no more fatal accidents. There had in fact been some pilot fatalities in the testing of the Macchi-Castoldi 72 seaplane fitted with the new Fiat AS6 engine, where the engine carburation system blew up when the machine was working up to maximum flight speed. This power unit was the ingenious brain-child of Zerbi, the head of Fiat's Engineering Department, and consisted of two separate 12 cylinder units mounted in tandem on a common crankcase. The two engines were not coupled together, the only mechanical connection between them being through the throttle lever and its linkage.

Each engine drove one half of a contra-rotating propeller through co-axial shafts, the rear one of which passed through the hollow reduction gear shaft of the front unit.

The complete engine had a total cylinder capacity of some 52 litres, or 15.4 litres larger than the Rolls'R' engine.

When I first looked through the Fiat test reports, the maximum power recorded for the Schneider Contest version of the engine was 2,400 bhp. In fact, while on test the sprint engine actually developed 2,850 bhp (but only for one minute) – which showed its potentiality if not its endurance. In fact, the development of this engine to give a reliable 2,850 hp took some weeks.

The AS6 test house was remote from the main plant and located at the Fiat airfield. The test equipment consisted of a double, contra-rotating, dynamometer or hydraulic brake, to take the complete power unit, the former being coupled directly to the main town water supply, with no capacity or header tank! When we were testing during the mid-morning there would often be a sharp increase in engine rpm, necessitating a shut down. This was attri- buted to the Turin housewives drawing their water supply for the midday meal!

I would sometimes turn up early at the test house, often to find nothing happening, after having agreed a programme the previous evening. The mechanics would bring me one or two sparking plugs completely oiled up, too oiled for normal fouling and with too fresh oil. I had arranged from the start with Fiat that they should use one of the two makes of plug we had previously cleared with Rolls on the 'R' engine, since I wished to be completely free of pre- ignition trouble from this source. But I found that an American plug company, who had previously supplied the plugs, was not going to let Fiat get away with our recommendation so easily. But I had my way.

My first job was to try and find why the engine carburation system fired back in flight.

Fiat had tried the forward, or ram air intake originated by Rolls- Royce. In theory, and in practice, this type of intake made use of the forward air speed of the aircraft to convert kinetic energy to pressure energy, thus increasing the pressure on the inlet side of the super- charger and reducing the pressure differential across the carburetter. But, in the Italian case, the intake was originally only fitted when the engine was installed in the aircraft and not on the engine test bed, and this proved to be the cause of the blow backs and engine failures.

In other words, in testing the engine on the bed without the forward air intake and with the maximum pressure differential across the carburetter, there was no trouble. Fitted with the forward intake in the machine, the ram increasing with speed, this pressure differential reduced and consequently weakened off the mixture which resulted in a severe and catastrophic backfire in the long induction pipe running the length of the double engine. I had deduced all this from the work at Rolls and got Fiat to instal a separate engine and blower in the test house, to simulate flight conditions and give a supply of ram air to the engine equivalent to the forward speed of the aircraft.

The rear unit, which drove the supercharger, naturally produced less power at its (adjustable pitch) propeller than the forward one, and the method of obtaining equal revolutions for each half of the contra-rotating propeller was by trial and error; the engine was run up on the ground and, according to the difference in speed between the two propellers, the pitch was adjusted.

While spending most of my time with Fiat in Turin, I also paid a

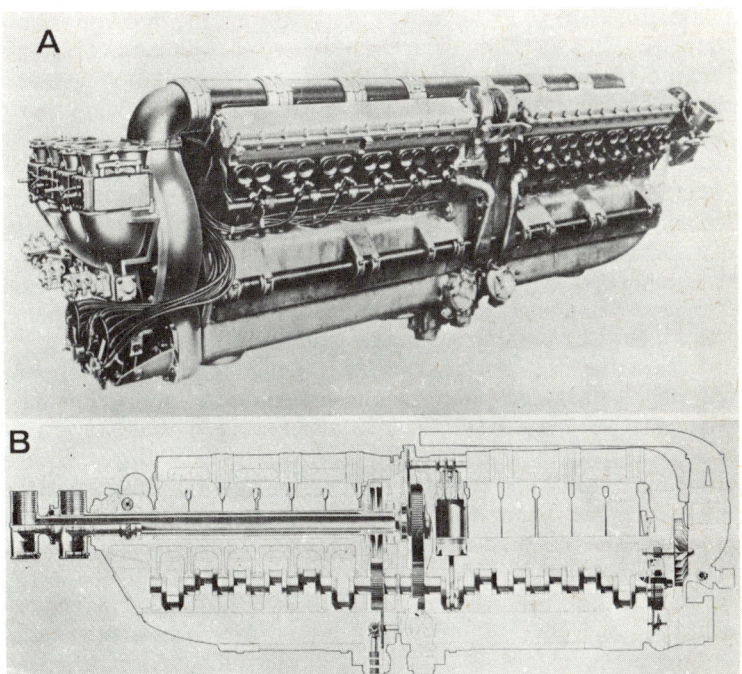

A. *The Fiat AS6 engine destined for the Macchi-Castoldi World Record seaplane. It had 24 cylinders in two separate and unconnected engines, each crankshaft driving one half of a contra-rotating propeller. Maximum power was 2,850 bhp.*
B. *Diagrammatic section of the AS6 engine showing the crankshafts and the inner and outer propeller shafts.* By permission of Fiat, Turin.

Record breaking Macchi-Castoldi 72 taking off from Lake Garda, 1933. By permission of Macchi-Castoldi, Fiat and the Italian High Speed Flight.

number of visits to Desenzano on Lake Garda, the operations base of the High Speed Flight, which was under the command of Colonel (later, General) Bernasconi. While there I met the Italian poet and patriot D'Annunzio who had been banished as a political nuisance by the Government and given a nice villa on the shores of Lake Garda with a converted motor torpedo boat as a yacht.

He used to invite us to his villa for dinner and always had attractive women to balance the guests. When departing at the end of the evening he presented the High Speed Flight officers with Italian wine and cheroots but always gave me a bottle of whisky and about 500 *Wild Woodbine* cigarettes – apparently under the impression that all Britishers smoked this particular brand.

Eventually, after much hard work and a great deal of help by the able engineers in charge of research and testing at Turin, Bona and Mancini, I managed to convince all concerned that the engine should first run for one hour at a full-throttle, 2,850 bhp, and then complete a cyclic test representing six runs up and down the three kilometre speed course, to prove performance and mechanical integrity. The fuel used was a blend of leaded gasoline, benzole and ethyl alcohol.

After successfully accomplishing these tests the flight engines were modified, check tested and returned to Desenzano for installation in the seaplanes. No further engine trouble was experienced and the Italians wrested the speed record from us in 1933 at 682.078 kph (422.888 mph), the pilot being Agello; later, it was further raised to 709.209 kph (440.709 mph) in 1934.

The latter record was only beaten in 1939 by Fritz Wendel of Germany when, on April 26th of that year, he flew the Messerschmitt Me-209VI at an average speed of 469.22 mph. This (land) machine was fitted with a special Daimler-Benz engine giving about 2,300 bhp. There was no radiator but an oversize water tank fitted internally took its place, the water being allowed to boil off during the relatively short flight.

On completion of this work at Fiat, I was presented by General Crocco to Mussolini at the Palazzo Venezia in Rome. I understood that he, (on Crocco's recommendation) wished to appoint me a Commendatore but, apparently, the British Government (Foreign Office) would not agree. Anyway, I heard no more about it.

14 British and European Military Aero Engine Developments

It was often said by Rolls at the time that the intensive work on the 'R' engine held back the further development of the Kestrel. There was probably some truth in this but it certainly brought along the Merlin for the Hurricane and Spitfire, which latter resulted more or less directly from Mitchell's racing seaplanes.

However, some years before our Schneider Trophy successes and the advent of the Hurricane and Spitfire, it was Richard Fairey who during the biplane era appreciated the aerodynamic advances represented by the clean-lined American Curtiss biplane, a seaplane version of which flown by a U.S. Navy pilot, Rittenhouse, had won the Schneider Trophy Contest at Bournemouth in the early 'twenties. Fairey went to the States in 1924, bought some of the very successful Curtiss D12, 400 bhp, engines and designed a clean two-seater fighter and reconnaissance machine, the Fox, around this motor. He then proposed taking a licence to build the D12, but the Air Ministry objected. One squadron (No. 12) of Fox biplanes fitted with the American built D12 was ordered and went into service with the Royal Air Force.

Fairey's initiative stirred things up and one result was that the Rolls Kestrel, an advanced design compared with the D12, eventually engined the Fairey Firefly, Hawker Hart and Fury machines. Later, Fairey put in hand the design of a vertical H type, liquid-cooled, poppet valve, double-crankshaft (unconnected) engine, each shaft driving one half of a contra-rotating propeller – rather on the lines of the Zerbi Fiat racing engine. Graham Forsyth had the design responsibility and engineering development of this 1,800 bhp engine, the Prince, and I helped him with advice on valve and piston design. But the engine received no support from the Air Ministry, perhaps rightly so in view of the number of engine manufacturers already existing, though, in my opinion, it had the makings of a reliable high-power unit.

Of the other manufacturers, Napier was not notably successful after the Lion became obsolete and they had built nothing worth-while (though taking the licence for the Junkers compression-ignition oil engine) until Halford designed for them the Rapier and Dagger engines which were also not very outstanding motors. It was not until he and his team designed the Sabre, a high-power fighter engine, that things started to look up for the Acton firm, more particularly since war was seen to be coming and rearmament had started. I have often wondered if Napier might not have done better had they taken on, or been given, the Fairey Prince, which was an easier production job than the Sabre and had good development potential. However, the 'ifs' and 'buts' are legion in this form of creative engineering.

In addition to almost continuous contacts with the British engine companies, I was similarly engaged with the American and European engine manufacturers, particularly with the airlines and the various foreign air forces who wished to take advantage of tetraethyl lead which could so eaily be stored in the event of war.

In January 1933 the Weimar Republic in Germany died and the Third Reich was born. Hitler became Chancellor and, ignoring the Versailles Treaty, openly signalled the rebirth of the German aircraft industry. Shortly afterwards, I was invited to give a talk to its aviation engineering leaders. This meeting was held in the old, and then defunct, German Colonial Office.

It will be recalled that, after the 1914–1918 War, the Germans were allowed to build only small engines, along the lines of the Gipsy in type and power (Argus and Hirth), but Junkers and others had set up design teams and prototype plants in Sweden and Russia to develop larger and more powerful engines.

It should be mentioned at this point that, in the 'twenties, it was General Hans von Seeckt[1], Chief of the Army Staff, and Admiral Erich Raeder, Chief of the Navy, who, despite the Versailles Treaty, kept a nucleus or cadre organization in being in their respective Services. These formed the basis for later expansion, particularly when Hitler assumed control. In Germany's original and post-war defence attitude, these two officers played an invaluable rôle. There was also a Captain Kraehe, who was the clandestine adviser on aviation and, later, his assistant, Captain Wilberg, directly seconded to von Seeckt, presumably to help organize a cadre within a cadre for the gradual build up of an air force (later, the Luftwaffe).

With the coming of the Third Reich there was no concealment of the fact that there would be a large and effective German aircraft industry and it was already being planned for rapid expansion.

After giving my talk, I returned home and went to see Major (Archie) Boyle in the Intelligence Section of the Air Ministry. I told him what I had done and said that while I could not obviously indulge in politics I was prepared to go easy in giving the Germans information. He was quick to say that I should continue my work in the normal manner and establish as friendly relations as possible, but he asked to be kept fully informed of all happenings and events. So I continued my advisory work in Germany and helped them with their engine test and development programme.

This work brought me into touch with Milch, whom I had met earlier in 1927 when he was in Lufthansa, and Ernst Udet. I had previously met Udet when he was an outstanding aerobatic pilot of international repute. Udet was Director General of procurement, embracing research and development and, under Milch, was responsible for production and getting into service the new aircraft types for the Luftwaffe. Later, I was to meet Goering, with Milch, on two or three occasions and, by official invitation, dined at Karinhall, his house some 60/70 kms from Berlin.

On one of my visits to Berlin I met a retired German naval captain, who was employed on liaison duties by one of the major oil companies. One evening, while dining together, he started reminiscing on the 1914–18 War and said he had been one of Christianson's pilots at the seaplane station at Zeebrugge on the Belgian coast. When I told him that I had operated in motor torpedo boats (CMBs) from Dunkirk, he mentioned that he had attacked one off Ostend early one morning.

I then told him of my own particular engagement with a Branden-

Curtiss D12 engine of 400 bhp.

Haus der Flieger
und
Clubhaus Rangsdorf

Ehrengaſtkarte

Nr. 141

für Herrn Ingenieur Banks

vom 11.X. bis 25.X.1938

Aero-Club von Deutſchland e. V.

Author's entry card to the Pilots' Club which was housed in the old Prussian Diet House, Berlin.

burg seaplane, and it transpired that he had been my opponent! As these things go, we became good friends and he helped me a lot in my contacts with various Government departments. He was particularly informative later when it became plain that the new Germany was out for trouble, since he was passionately against another war.

One evening in a taxi returning to the hotel after dinner in Berlin one of the party started singing Heinrich Heine's 'Lorelei':

> 'Ich weiss nicht was soll es bedeuten
> dass ich so traurig bin,
> ein Märchen aus uralten Zeiten
> das kommt mir nicht aus dem Sinn,
> die Luft ist kühl und es dunkelt
> und ruhig fliesst der Rhein
> der Gipfel des Berges funkelt
> im Abendsonnenschein.'

The taxi driver pulled in to the kerb and put his head inside and said 'You mustn't sing that. Heine was a Jew!'

In my earlier German visits I went to the Argus and Hirth engine companies who built small in-line, air-cooled engines of the DH

Gipsy size and type. These engines were fitted in the biplanes which were used as training machines for the nucleus of the newly-formed Luftwaffe.

At the time, since there were no U.K. or European manufacturing facilities for Ethyl, it was imported from the U.S.A. by our parent firm – Ethyl Gasoline. When the Germans indicated that they wished to build a lead manufacturing plant in their own country, this was at first resisted by us as uneconomic. Eventually, we had to agree to a joint manufacturing company formed between ourselves and I. G. Farben and called Ethyl Gm.b.H.

When this plant was built and operating, at Gapel, near Brandenberg, in the period of intensive German rearmament, we heard a rumour that they were contemplating building another lead plant, secretly. But we could get no confirmation of this. One day, however, we received, in our London office, the usual omnibus envelope containing the various departmental letters from I. G., and one of these was addressed to the German Air Ministry. Ray Bevan had it carefully steamed open and found reference to the new, clandestine plant to be built at Frose, ten miles to the north of the Harz mountains. Almost as soon as we had done this there was a telephone call from I. G. Farben asking us to send the whole envelope back unopened, which we did!

When members of the Air Ministry (R.A.F.) bombing or target committee used to visit us to get details and locations of these lead plants, it was always a source of curiosity or wonder to Ray Bevan and me that they were never bombed. In fact, after the War, the Russians literally tore down the Gapel and Frose plants, even to the lavatories, and had them shipped to the U.S.S.R.

By 1933 I was getting so personally occupied and involved in aero engine work that it was decided I should have an assistant who would concentrate upon the automotive side. We took on a very capable young research engineer, Hu Fossett, from the technical and test department of the then London General Omnibus Company (later L.P.T.B. and now London Transport), at Chiswick. Hu was to become Manager, Manufacturing, when the production of tetraethyl lead started in the U.K.

The decision to manufacture tetraethyl lead in the U.K. was initiated in 1935, when the Air Ministry approached us to find out what could be done to provide adequate supplies for the Royal Air Force in the event of an emergency. The American Neutrality Act made it doubtful that such supplies would be forthcoming from the U.S. in the event of a war in which she was not involved.

It was eventually agreed that Ethyl Export should form a British company in conjunction with Imperial Chemical Industries. This company would hold all the necessary drawings and know-how in the U.K., so that a lead manufacturing plant could be built and operated at short notice. Thus, on the 9th May, 1936, British Ethyl Corporation Limited was born, with a nominal capital of £100, the shares being owned 50 per cent by Ethyl Export and 50 per cent by I.C.I., the basic know-how coming from Ethyl Export and the operating staff from I.C.I.

In the meantime, stocks of Ethyl aviation (I-T) anti-knock fluid were imported into the U.K. and, in order to ensure a turnover, were also used for motor gasoline.

The use of lead in aviation fuel rapidly developed and there was also a steady, though small, increase in the distribution of leaded automobile fuel. By 1936, a number of European countries had premium grades of gasoline selling under the brand name of Ethyl. It was sold also in Australia, New Zealand and some of the South American countries.

During my earlier visits to the States I had heard of a material called Stellite, which it seemed could offer advantages as a valve seat facing material. Stellite No. 6, the grade used for valve facing, was composed roughly of 65% cobalt, 27% chromium, 4% Tungsten, 1.25% carbon, 2.7% silicon, with a hardness value of 45 Rockwell C at 127°C. It was then used for tipping the cutting edges of tools and also the blades of agricultural and earth moving machinery. Handforth, the metallurgist of Napier, early became interested in Stellite, but used a harder version of a similar material, Deloro, produced in Canada. I managed to interest the engine manufacturers in Stellite and, later, after some intensive development, it was possible to get a number of engines through Air Ministry Type Test with a clean (unburned) set of (Stellited) exhaust valves.

Sam Heron had been looking at Stellite but the American engine companies had yet to try it. When I told him of its use in the U.K., he really got started and personally learned how to apply it to a valve seat.

I always remember Sam and myself going to Hispano, in Paris, to show them how to apply Stellite. There were mostly women in the department and Sam, standing behind one, took hold of her wrists and guided her torch and the Stellite rod. She was large and fat, and a reputed Communist, but she enjoyed this embrace of Sam's who was very much amused and laughed uproariously.

Later, Brightray was developed for similar treatment of valves, by International Nickel. This material was composed, roughly, of 80 per cent nickel and 20 per cent chromium, and though less hard than Stellite had a superior hot corrosion resistance. Mostly, the valve head and seat, being hotter than the cylinder insert, were treated with Brightray and the insert with the harder Stellite.

On one occasion in Prague, I was cabled by my firm to go to

Bucharest, since the Rumanian Air Force wished to see me urgently. I arrived by air and went to the Athenée Palace Hotel where I found a message from the Air Minister inviting me to his apartment for luncheon the following day which was a Saturday.

The apartment, incidentally, was opposite that of Mme Lupescu, later to be the (morganatic) wife of King Carol. It was guarded by plain-clothes police, who asked me my business before letting me go to the Air Minister's apartment.

Towards the end of the meal, at the coffee stage, the Minister explained that there was to be a meeting at the Air Ministry on the following Monday to decide upon the use of leaded (87 octane) fuel for the Air Force. 'Of course', he said, 'there may be ways by which we could come to an understanding'. I went all oriental and sipped my coffee and after a moment explained to him that we did not do business that way. I further added, as tactfully as I could, that the whole of the Rumanian Air Force did not use in one year as much fuel as two or three Great West Road service stations dispensed on a Saturday afternoon in summer. He merely smiled and shrugged and didn't say anything.

In the event, the meeting on Monday decided to go over to leaded fuel. There was, in fact, little alternative fuel for their engines.

In those days, a year or two before the Second World War, Rumania was an amusing but at the same time exasperating country. There was currency control upon entering and one was instructed that travellers cheques could only be cashed at the Bank of Rumania at the current rate of exchange of L400 to the Pound Sterling. Few took any notice of this and since most of my business was with the oil refineries at Ploeste I merely asked them to cash my requirements, which they did at L800/£ – double the official exchange rate and halving the value of their money at a stroke!

On occasion, we used to spend the odd evenings at various 'Nacht Lokale' with some of the tougher oil refinery people. There were quite often rough-houses, particularly involving Rumanian Army officers. These cavalry types were alleged to wear corsets, and they certainly made up their faces with rouge and powder. One evening I I was with two Americans from a Ploeste refinery and one, a little drunk, started taunting some officers at another table. In a moment tables were upturned, swords drawn and bottles and glasses thrown. The other American and myself grabbed our friend and hauled him out in double-quick time.

I met many interesting people in Rumania, particularly Peter Carp, the capable engineer in charge of the aero engine side of the State aircraft factory, I.A.R. (Industria Aeronautica Romana), at Brasov. This company had licences for producing the Polish PZL, gull wing fighter and the French Gnôme-Rhône engines.

Peter got himself into trouble and was imprisoned in a fortress, becoming seriously ill and nearly dying owing to the dampness of his cell. Both Devereux (of High Duty Alloys) and myself visited him there and tried to get him released, which came about soon after our efforts but not necessarily because of them.

The trouble started when the Gnôme-Rhône engines they were building were giving high lubricating oil consumption. Peter, with his almost Germanic viewpoint, decided that the important thing at the time was to get the engines cleared and installed in the airframes already awaiting them, then to concentrate upon the oil consumption problems at leisure. So he rigged up the test bed with a topping up pipe to maintain, by a remotely controlled electric motor, the oil level at a predetermined rate of consumption during the engine's Type Test running. Unfortunately, someone in the works, who presumably he had annoyed, informed on him and the official inspection organization then took over and prosecuted. For some reason, not quite clear even at the time, he was court-martialled and sentenced by a military court to a year's detention in the fortress. Possibly a court-martial was held rather than a civil trial because the offence was committed at a State factory on equipment intended for the Air Force.

Peter joined the largest tractor manufacturer in Rumania in the Second World War from which the Germans ordered half-track vehicles. But both he and his employer generally created so many difficulties and delayed production so much that the output was reduced to less than half. Because of this we were able, after the War, to get Peter back some of his money that had been confiscated in U.K. because he was an enemy alien.

I also visited PZL in Poland a number of times and the Polish LOT airline. PZL had taken the licence for the Bristol Pegasus, and it was Zenkowski, head of the engine plant, who finally convinced Roy Fedden and Bristol that their floating big-end bush was a snare and delusion and caused them to revert to a fixed bush. I mention this only to show how important it is to have good licencees with capable engineering and development departments who can help the parent company improve its product.

15 *Fuel Developments in the Biplane to Monoplane Era*

In the early and middle 'thirties some important happenings occurred that were profoundly to influence World events. After the final Schneider Trophy Contest in 1931 there came, in November 1935, the prototype Hawker Hurricane, to be followed by the Supermarine Spitfire in March 1936.

These two aircraft, which were to form the backbone of the Royal Air Force's fighter strength and be our saviours in 1940, were fitted with the original Rolls-Royce Merlin engine, developed from the PV12 (private venture) motor. They broke the biplane philosophy of the Air Staff, who had largely followed Trenchard's precept that it would be the bomber that decided the next war and bring an enemy to its knees. How wrong this proved to be when the War came and Germany was saturated by bombing, but did not give up primarily because of it; although Albert Speer has since the war acknowledged that, had it continued, the bombing would have finished the war for Germany.

In Germany, engine development was also rapidly progressing, with the four principal companies engaged – Daimler-Benz, Junkers, B.M.W. and Siemens-Halske (later, Bramo-Hafnir). The two former built inverted, 12-cylinder, liquid-cooled engines and the two latter, air-cooled radials. The liquid-cooled German engines were of somewhat larger cylinder capacity than the Merlin, around 34 litres compared with 27 litres of the British engine.

In the case of the Daimler-Benz, it was also designed to take a *moteur canon*, as the French originators termed it – a 20 mm gun fitted in the V formed by the cylinder blocks and firing through the hollow shaft of the propeller reduction gear.

This arrangement prevented the fitting of the supercharger in the normal position at the rear of the engine, so it was placed edge on in the fore and aft plane on the side of the engine and driven at 90 degrees to the crankshaft. In this position, it was not practicable to fit the carburettor to the inlet side of the supercharger in the normal way, since it would have produced a large bulge in the engine cowling, so a pressure carburettor was fitted between the cylinder blocks. But German carburettor development was somewhat behind our own and a pressure carburettor was in any case difficult to make work well. So they discarded it in favour of the multi-point injection system spraying directly into the cylinders.

The fact that the Germans did not persevere with their carburettor development was to prove a blessing to them. The British carburettor with its variable datum boost control was undoubtedly a very refined piece of mechanism and equipment, but was later to prove a sore trial to fighter pilots in World War 2. When the nose of a machine was put down to dive on an enemy, this resulted in negative 'g' on the float chamber and caused a rich mixture cut of the engine – allowing the enemy machine to get away.

The first of the long series of Merlin engines was not too good on exhaust valve condition and life. In fact, it had difficulty in passing the full official, 100 hour, Air Ministry Type Test. This was mainly due to the cylinder head design having a sloping or angled surface, known as the ramp head. Apparently, the head structure was insufficiently strong or rigid and became distorted. The problem was aggravated by differential mixture distribution. Anyway, the engine rarely got beyond 70 hours on Type Test before exhaust valve burning occurred and terminated the test.

The French at this time depended mainly upon the engine firms of Hispano-Suiza and Soc.Gnôme et Rhône, the former building a 60° V-type, liquid-cooled, 12-cylinder engine and the latter two-valve, air-cooled radials. From their experience of building the first Bristol *Jupiter* under licence, in 1920, Gnôme had gone along on their own lines and turned to the classic two-valve cylinder head, designed by Roger Ninnes who came from Bristol.

From the excellent 90 degree V-type, 8-cylinder, water-cooled, engine of 180 and then 250-300 bhp, in the 1914–1918 War, and the unsupercharged 60 degree V 12-cylinder engines in the early post-war years, their later engines immediately before the Second World War period did not produce the specific power output or reliability of the British or German counterparts. The French engines had inefficient superchargers, tubular connecting rods, somewhat skinny crankshafts and weak crankcases.

When it came to supercharger design, Marc Birkigt's good engineering sense seemed to desert him. His supercharger looked like an overgrown water pump, with the casing little larger than the rotor. It had no diffusers and the temperature rise was phenomenal. It even burned the paint off the outside casing at very low blower pressure ratio. British supercharger experience had shown that, without diffusers, the inside diameter of the supercharger casing should not be less than twice that of the rotor, giving the mixture a chance to straighten itself from the turbulent condition before passing to the induction pipe proper.

This weakness in supercharger design provided fertile ground for Szydlowski and Planiol when they entered into the proprietary supercharger field with their aerodynamically designed and variable inlet stator blower, which was fitted to a number of French engines before the Second World War.

Szydlowski is now the presiding genius of Turbomeca. Planiol went to the U.S.A., but died there some years ago.

Italian engine manufacturers, like Rolls-Royce in the 1914–18 War, had largely copied German engine practice with the welded steel-jacketed cylinder, and did not show great originality in the design of high output aero engines between the two World Wars. This was always surprising since they were good mechanical engineers and their automobile engines were outstanding in many cases. But they were to build German engines in the coming war.

The U.S.A., after the quantity-produced Liberty engine of the First World War, did little in the liquid-cooled engine field afterwards, with the brilliant exception, and conception, of the 400-bhp Curtiss D12 engine in the 'twenties. It was later followed by the larger Curtiss Conqueror but, after that, apart from some interesting work by Continental and Lycoming on very high output liquid-cooled cylinders, under contract from Wright Field, the liquid-cooled engine was dropped until the Allison came in 1939. The U.S.A.A.F. had encouraged Lycoming with contracts for the very high duty (Hyper) single cylinder development, envisaging a 'flat' 12 cylinder engine of 1200 cu in capacity and 1000/1200 bhp. But its development was stopped due to the lack of facilities and funds at that time. This interesting cylinder development emanated from the brains of (the late) Val Cronstedt, chief engineer of Lycoming, and S. D. Heron.

The Americans concentrated mainly on the air-cooled radial, produced by Pratt & Whitney and Wright Aeronautical. Both made outstanding contributions to military and commercial aviation.

To America should go the credit of discovering the only practical anti-knock compound, tetraethyl lead. Largely through the experimental work done by the U.S. Army Air Corps (now U.S.A.F.) at Wright Field, in collaboration with the large oil companies (Esso and Shell), the Americans also gave us quantity production of high anti-knock fuel. We, in the U.K., were to benefit from this in the middle and latter 'thirties and in the War, first by supplies of 87 octane and then 100 octane gasoline.

It is interesting to note here that the unsupercharged Liberty engine of the First World War, having the same cylinder capacity as the Merlin (27 litres), produced 400 bhp on a straight-run on gasoline (petrol) of about 58 Octane Number, whereas the supercharged Merlin in the Second World War gave over 2,000 bhp on 150 grade (Performance Number) fuel.[1] A little less than half of this increase could be credited to the improved fuel and the balance to supercharging (with after-cooling) permitted by such fuel.

The unsupercharged racing Napier Lion in the Supermarine S.5 float seaplane which won the 1927 Schneider Trophy Contest at Venice gave nearly 900 bhp at 10:1 compression ratio on a straight-run light Rumanian gasoline having a lead concentration of 7 ccs per Imperial gallon. Almost as important to engineers, this engine was subjected to an officially observed test, with Ricardo as official observer, and recorded a specific fuel consumption of 0.32 lb/bhp/hr – practically the theoretical possible for that compression ratio.

In the 'thirties I assisted Hawker's test pilots, 'George' (P. W. S.) Bulman and Philip Lucas, in the demonstration flights they gave for potential military customers. This was done in a rather simple or crude way, by meeting them at the appointed time and place anywhere in Europe and, with my one litre container of 1T Ethyl Fluid and rubber surgical gloves, doping the fuel directly in the aircraft fuel tank, so that they could fly at full throttle throughout the demonstration without having to worry about the throttle gate.

While such demonstrations might at first have appeared unrealistic in view of the quality of fuel generally available at the time, they had the very practical value of showing how improved fuel could permit increased fighter performance, and proved very important later in showing what was possible when 87 and 100 octane fuel became generally available.

I remember my first meeting with Air Marshal Sir Hugh Dowding (the late Lord Dowding) when he was at the Air Ministry, in 1935, as Air Member for Research and Development (A.M.R.D.). The Ministry had approached us to find out what could be done about providing adequate supplies of tetraethyl lead for Royal Air Force fuel stocks in the event of war, and Ray Bevan and I went along to see Dowding. When we were ushered into his room by his P.A. he was reading some paper or report, and continued to do so for a minute or two without looking up. To me, a minute or more of silence and inattention when going into someone's room savours of rudeness and is certainly discomforting to the visitor. Eventually, 'Stuffy' looked up and, without apology, started firing questions at us. He was, in fact, quite courteous as the meeting progressed.

I was later to meet him in the War, when he was A.O.C.-in-C. Fighter Command, both at the start and in the thick of the Battle of Britain; and though I was a junior officer (Squadron Leader), he was always extremely nice to me and most anxious to get any help to keep his squadrons flying and avoid unnecessary delays or engine troubles that could cause accidents to his pilots. At that time, I was directly concerned with sparking plug behaviour which, with the newly introduced aluminium oxide insulators and fine wire electrodes, was giving rise to some anxiety due to uncertain operation and a high (engine) failure rate. I had originally brought back from Germany the first examples of this type of plug, developed by Siemens.

One day in 1936 a friend of mine in the continental shipping business, who knew and saw Churchill from time to time, asked me if I would like to meet the great man. Naturally, I said 'Yes'. Churchill was then in the political wilderness but had been quite vocal in warning the Government and the nation of the dangers of the Nazi movement to Europe and to the world. My friend thought I might confirm for him some of his views. A week later we went to his flat off Ashley Gardens and were ushered into the living room. Churchill came in and I was presented to him. He then strode up and down and gave a very interesting dissertation on the European situation. After about forty-five minutes, he said he was pleased I had come along. When we got outside my friend said, 'Wasn't that marvellous'! I said, 'Yes, it certainly was but I thought the idea was that I should confirm to him his impressions of Germany'. I hadn't uttered a word.

*   *   *

During one of my frequent visits to Germany, in 1936, in connection with the engine tests I had initiated and was monitoring, I was invited by Udet to see the famous aviatrix, Hanna Reitsch, fly the twin-rotor Focke-Achgelis helicopter in the huge Deutschland-halle in Berlin. The engine noise was considerable, but the demonstration of a full-scale helicopter that could actually fly and be manoeuvred was most impressive. I met her after the demonstration, but only saw her fleetingly, later, at various Government receptions. She was an ardent admirer and supporter of Hitler.

Surprisingly, I met her again in 1965, at a reception given in my honour in Accra, Ghana, of all places. She was then running a gliding school in the northern part of Ghana.

It was Hanna Reitsch who, in the War, flew the V1 flying bomb to check its actual performance and range. Later, at the War's finish, she flew into Berlin to try and persuade Hitler to fly out with her. It was rumoured that she was prepared to die with him in the Bunker, but he ordered her away.

In the between wars period, the Lilienthal Gesellschaft was formed in Germany. This was the counterpart of our Royal Aeronautical Society and named after the German aeronautical experimenter and glider designer and pilot, Otto Lilienthal (1848–1896).

Wolseley Viper engine of 200-300 bhp (Wolseley-Hispano, 90°, V8, direct drive unit). By permission of the Science Museum, London.

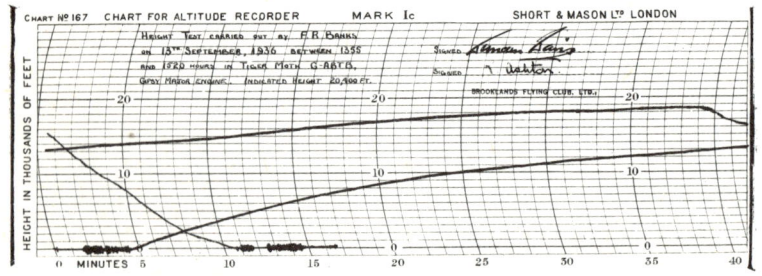

Barograph chart of 'height record' at Brooklands by author in Tiger Moth.

It succeeded the Vereinigung für Luftfahrtforschung which had come into being in 1933. The Germans, to gain further knowledge and information for their rapidly growing aircraft industry, organized, through the Gesellschaft, a number of international get-togethers, inviting foreign aviation personalities to attend and to read papers. Fedden was one, since they wished to know about the sleeve valve engine on which he was then doing a great deal of intensive development at Bristol. On one occasion, after a meeting a select few of us were taken to a small private room and presented to Hitler by Milch. He shook us by the hand but hardly said a word, just nodded his head as we were introduced.

My other visits in Europe included trips to Poland on Ethyl blending operations at the oil refinery at Lwow (originally Lemberg under previous Russian rule).

I helped the Walter Aero Engine Company of Prague to improve their cylinder head and valve performance. Teny Kumpera, the head of Walter, when he saw what was coming in the Sudetenland and the probable takeover of Czechoslovakia by Hitler, suddenly arrived in London with two cars, one driven by his father. Bolted to the bumpers and chassis were solid bars of gold. They had driven across Europe and over many frontiers without these being detected!

He went straight to our mutual friend Devereux (High Duty Alloys) at Slough and asked him to dispose of the gold and turn it into useful currency! This, naturally, presented Dev. with an embarrassing situation, but he managed to get it settled satisfactorily with the Bank of England.

In the early 'thirties, when the biplane Hawker Fury was in service in the R.A.F., I flew over with 'George' Bulman, chief test pilot of Hawker, in their famous Hart 'MR' on the occasion of the International Air Meeting at Brussels. We were there to look after No. 43 (Fury) Squadrons from Tangmere, who were to give a series of flying displays, and I was to dope their fuel so that they could use unlimited throttle. The names of two pilots in '43', Mermigan and Hawtrey (nephew of Charles Hawtrey the famous actor) will be known to those of their contemporaries who may read this.

Another year, at the same International meeting at Brussels, I flew over with Tommy Rose, the Anglo-American Oil Company (Esso) pilot, in an Avro Sports Avian of which I had taken delivery for Anglo-American, at Manchester from the hands of the works manager, Roy Dobson (later their managing director, Sir Roy Dobson). When we were over Brussels airfield Tommy commenced his usual 'shoot up', which was tolerated in those days. When we dived on the field I felt a distinct shudder through the machine. After two dives and quite high 'g' pull outs, Tommy also got the message but thought he had gone over the edge on speed. However, later, when we were putting the machine away at night and pulled out the lower starboard wing bolt to fold it, the whole wing folded back on its own! The upper bolt was apparently in the 'locked' position but had overridden the centre section fitting. This led to a modification.

In 1934 I took delivery of one of the first of the DH Rapides on behalf of Ethyl Export, with Tommy Lucke, a Short Service Commission pilot we had engaged upon his demobilization from the R.A.F.

On one of the early trips in this machine, a flight to Prague, we took off from Brooklands on a very stormy and blustery day, landing at Lympne to clear Customs and then flying on to Cologne. The weather had deteriorated further, with low cloud and heavy rain so, after refuelling, I decided we should stay in Cologne for the night. But just as we were about to get the machine into a hangar, an Imperial Airways DH86 arrived from Leipzig, our next staging point, and the pilot told us that the clouds were well above the Harz mountains and it 'Wasn't too bad.'

We decided to go on to Leipzig, despite the early darkness of a winter's day. I ought to have known better since, at that time, I had more knowledge of European flying than Lucke.

After we had been flying for nearly an hour, I looked up from the book I was reading, as it appeared to be getting quickly darker, and saw Tommy Lucke looking out from side to side of the cockpit as if he didn't know where he was. He didn't! The radio had apparently packed up and he had flown off his strip map. I told him to land immediately on any flat spot he could find. But there were no flat spots and we were still over the undulating Harz mountains. It got dark and flying at only about 500 feet above the hills, we saw the lights of a town. I went far back in the cabin, nearest the tail, after telling Tommy to motor it down on anything that looked flat near the town.

I woke up sometime later, having been flung from one end of the cabin to the bulkhead behind the cockpit, narrowly missing the sharp edges of the radio set bolted to it. Apparently, Tommy had 'landed' about 20 feet high, above a ploughed field, and the machine finished on its back. Having been knocked unconscious and with blood pouring out of a deep scalp wound I didn't realize, when I woke up, that we were upside down; but was eventually fully aware of the sound of dripping outside, which I decided was petrol. This made me collect my thoughts very quickly though I did not remember the actual crash and, stumbling about on the cabin roof, finally found the door, which I could not open. Tommy, miraculously uninjured, had scrambled out of a broken window in the cockpit. With me feebly kicking the door, and he pulling more effectively, it opened and I fell out into the mud. The machine, luckily, did not catch fire.

The inhabitants of Halberstadt, the town, hearing the machine circling around, soon got to us. I was taken to hospital, where my cracked skull was very capably stitched up and I was given an anti-tetanus injection.

Tommy Lucke then rang Con, my wife, and told her that we had stopped the night at Halberstadt and that 'Rod was very tired and had a headache and gone to bed'. She didn't believe a word of it, but I was able to ring her later the next day and explain things.

The official (German) fuss came the next day, since Hitler was then in power and the local Nazi hierarchy came to question me in hospital but were held off for a further day by the doctor. Eventually, a German Esso aviation friend of mine, Emil Kropf, Colonel in the German Air Force Reserve, was telephoned and he came to assure the local authorities that I was not a spy.

It was a week before I left hospital to fly home. In this time the Halberstadt flying and gliding club members did a very good job of dismantling the Rapide and packing it for transport to Hatfield.

After this incident, and when it had been repaired, Tommy Lucke flew the machine all round Europe, the Middle East and Africa, with company members, without bending anything.

Eventually, we got rid of it and I think it took some part in gun running in the Spanish War.

*   *   *

The Front Populaire Government came to power in France in 1936, with the Socialist, Léon Blum, as Prime Minister. He appointed as Air Minister Pierre Cot, who immediately planned to nationalize the French aircraft industry. Cot had already been Air Minister at the time of the decree of the 1st April, 1933, when the French air arm was named Armée de l'Air. He left in 1934 when the Armée de l'Air became independent of the French Army and the post of Air Minister was taken by General Denain. Denain wasted little time in setting a new pattern for French military aviation, obviously being influenced by the military growth of Nazi Germany and Fascist Italy and the threat of a blitzkrieg. The French knew about this form of war since a French mission had been to Russia in 1933, at Soviet invitation, and participated in an early form of blitzkrieg involving combined ground and air exercises with close support aircraft and tanks. It would also not be surprising if this early exercise had been dreamed up by the German generals who helped to train the Russian Army after the signing of the Treaty of Brest-Litovsk.

General Denain's proposals were for a strategic bomber and tactical fighter force but, apparently, General Weygand, the Army Chief of Staff, would not agree, though the French Navy supported

Denain. Presumably, Weygand couldn't get used to the fact that the Air Force had been parted from the Army!

I got to know the chief of the Service Technique well at that time, Ingénieur-Général Albert Cacot. I liked him and thought he had knowledge and good sense.

I have given this brief historical sketch in order to emphasize and lead up to Pierre Cot's influence and nationalization plan when he returned as Air Minister in 1936. Indeed, before General Denain's plans were in full effect, here was Pierre Cot upsetting the applecart and pressing on with nationalization, despite, or in face of, the serious events again threatening European peace, with the Wehrmacht marching into the demilitarized zone of the Rhineland and, a few months later, the start of the Spanish Civil War.

The date of 11th August, 1936, was therefore very significant since this was the day when the French aircraft industry was declared nationalized. My impression at the time and since was that this decision, which was only half-baked, largely sealed the fate of the Armée de l'Air in the coming war by its interference with planning and the delays it caused in ordering effective aircraft that should have been tested and in full production in the three years preceding the declaration of War. We in the U.K. were not so different inasmuch as we had, up to 1936, no worthwhile fighters until the Hurricane and Spitfire came along, and were not to have an effective bomber until well into the war period, when the four-engine aircraft came along. But, at least, we did not have nationalization to contend with and hinder what was ordered and under development, with our production planning becoming fairly well organized and established.

The French Chief of Air Staff at that time was General Féquant and he was succeeded by General Vuillemin.

It was not generally known that, in January 1938, a French mission composed of Captain Rozanoff, a test pilot (later killed at an official post-war demonstration at Melun-Villaroche), Commandant de Bricy, an armament specialist, and Ingénieur en Chef du Merle were invited by the Spanish Republicans to inspect and fly two of the latest German aircraft falling into their hands, i.e. a Heinkel He.111 and a Messerschmitt Bf.109D. Later in 1938 General Vuillemin was invited by General Milch, the German Secretary of State for Air, to visit and inspect the Luftwaffe and see some of the German aircraft factories. When he returned to Paris, Vuillemin asked my good friend and colleague Colonel George Thenault (Commandant of the famous Lafayette Squadron in World War I) if he would ask me to come over for a talk. We met in Vuillemin's office and he told me that he had seen 'hundreds of Bf.109s on one field' and what could I suggest that would speed up things in France. As a matter of fact, the French Air Attaché in

Berlin, Paul Stehlin (the late General Stehlin), had actually got drawings of the Bf.109, with details of its performance which, incidentally, proved to be some 100 kms/hr (62 mph) faster than the Morane 405 (the pre-production version of the 406) and one of the best French fighters at the time. The performance figures of the 109 were also probably confirmed by Rozanoff's testing of this machine in Spain.

I could only suggest to the C.A.S. that, as a preliminary, he might persuade Hispano to invite Rolls-Royce's help in developing the HS77 engine to produce 1,000 bhp from the 850 bhp it was then giving in the Morane. But Vuillemin didn't think this would be practicable since it might offend the *amour propre* of Hispano's technical chief, Marc Birkigt. I thought this could well be so!

The trouble with Hispano was that they were backward in super-charger design, whereas Rolls already had considerable background and experience of successfully supercharging their engines. But the importance of Rolls' help would have been to advise on the re-design and strengthening of the Hispano engine to take the increased stresses and heat-loading imposed by higher supercharging. At that time, Rolls were in the throes of developing the Merlin and had yet to get the power and reliability needed, but I am sure they could have given material help to Hispano if it had been politically possible to arrange a get-together.

I also advised Vuillemin to obtain manufacturing licenses for British or American air-cooled radial engines. He was depressed because of the effects of nationalization; even Pierre Cot, who had started it all, finally resigned because *he* couldn't get the Government to agree the necessary funds for expansion. He was succeeded by Guy La Chambre.

The members of the French mission to Spain were impressed by the sturdy practicability of the Russian fighters, though they did not have the performance of the German equipment. This was to become apparent later when Germany invaded Russia and we supplied Hurricanes to help the Russians. The Russians did, how-ever, make a stupendous effort during the War and subsequently brought out some quite good tactical fighters and bombers that were effective against a weakened German Army and Luftwaffe.

So, with the Germans re-arming and becoming more threatening to all, I continued my European visits, reporting regularly to Archie Boyle at the Air Ministry. I remember, for instance, being in Prague at one time and receiving a cable from Ray Bevan asking me to go to Budapest as the Standard Vacuum Oil Company there wished to see me. Arriving in the middle of the afternoon, I saw Standard's managing director who explained the position they were in *vis-à-vis* the supply of fuel and oil to the Hungarian armed services.

Under the Treaty of Trianon (Versailles), the Hungarians were not supposed to have an air force, but Mussolini had given them aircraft and the Germans were promising to demonstrate and give them some fighters. Standard were not, however, supplying direct. This was done by a very attractive girl who had been with them and then left to set up her own oil agency, buying supplies from Standard. I was not allowed then, in 1936 and 1937, to go to the War Department, so it was arranged that I should meet the Chief of Staff at the girl's apartment in Buda, the old town across the Danube. There, I had tea with her and the C. of S., and discussed with him his technical problems. We then went out to dine and dance at a night club. I made several of these visits and was caught in Budapest on the eve of Chamberlain's last Munich (1938) talk with Hitler. Dining that evening with the head of Standard at his home, which was on a hill a little way from the centre of Pesth, we were beset by the noise of military transports hauling up A.A. guns, ammunition and search-lights, temporarily dumping ammunition from overloaded lorries by the side of the road. Back in my hotel at midnight, there was the usual crop of rumours; 'Yes, there would be war'. 'No, the Foreign Secretary has just been in and says there will be no war'. That day, before the final Munich meeting, most of the airlines, Imperial Airways and Czechoslovakian Airlines, had cancelled their services and I was debating how I could avoid being caught up in it all and how, if war came, to escape internment. I then remembered that K.L.M. had flights from Holland to the Dutch East Indies and went to see their Budapest agent whom I knew quite well. He said there was an aircraft on its return flight due to arrive that evening in Athens, and he reserved me the only available seat. So, on the day of Munich, I was able to fly to Amsterdam, refuelling in Germany on the way!

I went back to Middle Europe once or twice afterwards and, in 1939, was allowed to enter the Hungarian War Department H.Q. to see the Chief of Staff in his lair. In the meantime, Goering or Milch had already been to Budapest with Udet, who demonstrated some aircraft.

Back in October 1937, Milch and Udet, with Generals Stumpff and Wenninger, came to the U.K. at the invitation of the Air Ministry to be shown our latest equipment. They visited Odiham, Cranwell, Mildenhall, Halton and Hornchurch, but our aviation press, strangely, were not encouraged to accompany the mission. This visit was ostensibly in return for a visit by Air Vice-Marshal Courtney and Air Commodore Evill to Germany in 1937.

In July 1939 I was in San Francisco attending the Society of Automotive Engineers World Automotive Engineering Congress, and also presenting a paper. At the last day's session, when we were at the cocktail party preceding the closing banquet, I was talking with Fred Zeder, President of the Chrysler Corporation, the Guest Speaker, who asked me if there would be a war and I replied, 'Yes, unfortunately, I think so'. In his Address at the end of the dinner, Zeder said, 'Rod Banks and Leslie Callingham (Shell) think there will be a War in Europe. But I tell you all that not one American boy will be there'. Ultimately, with America in the War, his two sons were in the Services. It was at this Summer Meeting that I met Doodie and Jack Macauley. Jack was with Chrysler on experimental work and was getting over a bad car accident. He had hit an unlit farm cart one night and permanently damaged his leg, though he could still drive well. They were my friends for many years. Doodie and Jack, who retired to Alexandria in Virginia, across the Potomac from Washington, D.C., are sadly both gone. Jack went to Pratt & Whitney during the War as an experimental engineer and was developing a free-piston aero engine exhausting to a turbine for high economy propulsion. He afterwards joined the Ethyl Corporation, where we were in close touch over some years. He then retired and went to the Department of Defense as an Assistant Secretary R. & D. for a period until his final retirement. When I returned from San Francisco in 1939 I went to France, Germany and Italy. In Germany I dined with Udet and afterwards went to his apartment where it was the usual thing to shoot pistols at a sandbox target. There we were joined by an S.S. officer from another apartment and also a Hohenzollern princeling in the Brown Shirts. Udet in his office the next day seemed to be rather depressed about the state of things and didn't see how war could be averted. Yet previously, in June, before my American visit, I was dining with our faithful German represent-ative, Richard Morgenthaler, and a D.V.L. (German R.A.E. equivalent) friend, in an old established restaurant, the Gebrüder Habel, just off the Unter den Linden, when who should come in but the Chief of the Berlin Police with the Chief of Police of Danzig, and their staffs! Some of the restaurant's guests didn't appear to like what they saw, and indeed were most surprised to see the Poles there after all that was going on including blasts from Goebbels' propa-ganda machine every day on the 'iniquities of the Polish corridor' etc.

After that visit in July, I saw the last of Germany until the end of the War.

During my last American visit before the War, I was at Wright Field discussing engine and fuel matters with Colonel Eddie Page, Sam Heron, Opie Chenoweth and others when 'Dutch' Kindelberger flew in on board his prototype North American B25 (Mitchell) twin-engine support bomber. The next day an Army test pilot took it up and crashed it! Poor 'Dutch' was very upset and since it was the weekend, a number of us organised a barbecue for him in the wooded area behind the Field. We all tried to make him cheerful but I'm afraid our efforts only made more obvious his loss. However, the Mitchell eventually proved to be a good performer when, later, it went into active service in large numbers.

Colonel Eddie Page, who followed a succession of capable administrative engineers (Major George E. A. Hallett, E. T. Jones, Captain T. E. Tillinghast) at McCook and Wright Fields, was in charge of the Power Plant section at Wright for some years; and Sam Heron was there under him before joining the Ethyl Corporation. Eddie was a true Southern gentleman in every way and he knew well how to handle Sam.

Con and I were very touched when, at the beginning of the War, with children being evacuated to Canada and the U.S., Sam wrote to us to say that Eddie Page and his wife, who were childless, had offered to have our two children, Christine and Annrobin, for the duration and Sam would pay all expenses. This was extraordinarily generous and kind on the part of all three. But we declined.

---

[1] By definition, 100 per cent iso-octane represents 100 Octane Number, i.e. Performance Number (PN) 100. But at anti-knock values above 100, the PN scale is determined by adding tetraethyl lead to iso-octane, to give, in a supercharged single cylinder unit, a power increase of, say, 130 per cent of the power possible on 100 per cent iso-octane, i.e. PN 130.

16 *The Second World War*

In August 1939 I was on holiday with my family on the Isle of Arran off the Clyde Coast. Before we left London I had been to the Air Ministry and talked with various friends there. I also saw Tedder, who was then Air Vice-Marshal and Director General of Research and Development. We discussed what I might do if and when war started, and I said that since the First World War I had acquired a certain specialist knowledge and would like to put it properly to use. Tedder left it that I should contact him as soon as things happened.

Towards the end of the month, and approaching the end of our holiday, events appeared inevitably to point to war, so I decided to return to London leaving the family in Arran until we saw more clearly what was going to happen. In the event, they didn't come back to London but stayed in Scotland with my wife's mother until 1943.

I saw Tedder who took me to see Air Chief Marshal Sir Wilfred Freeman, Air Member for Production and Research. After a short talk, Freeman suggested that I join the Royal Air Force and they would then send me to A. & A.E.E. (Aircraft & Armament Experimental Establishment) at Boscombe Down. This was on the Wednesday and we were at war with Germany on the Sunday following. I was given my commission as a Pilot Officer, Acting Flt. Lieutenant, in the Administrative & Special Duties Branch of the R.A.F.V.R., the following week. But no sooner had I been commissioned than I received an urgent request to come to Paris from the French Air Ministry at the instigation of the Chief of Air Staff, General Vuillemin.

Tedder, to whom I was then immediately responsible, agreed that I go but we both thought that I should be in mufti for my dealings with Ministers and Generals.

I was in France for nearly a month, visiting the aero engine manufacturers and having many discussions with the Air Ministry's technical people (*Service Technique*) and, of course, General Vuillemin. It was then that I became worried about the French situation, since those outside the official circle, and even some in it, seemed to have a rather *laissez faire* attitude towards the war. Admittedly, it was only the start of what was soon to be known as the 'phoney war', but there seemed to be a lack of purpose or energetic prosecution of the more obvious wartime needs. All this modified an opinion I had previously held when, not knowing anything of the French Army, I thought it might give a good account of itself. Now, I was worried since some of the thinking people I met were either apprehensive or didn't seem to take in the seriousness of another war with Germany – or perhaps they did! When I returned home, I went immediately to see Archie Boyle (Air Ministry Intelligence) and told him of my fears. He said, 'Rod, I have always told you that, after her battering and losses in 1914–18, France would not be able to withstand another full-scale war'.

I advised General Vuillemin to get British or American help to speed French engine development, preferably British at that time. He wished however to buy time and get engines from the U.K. and U.S., but also queried whether or not to acquire complete aircraft since they had already done a deal for a number of Curtiss Hawk fighters, which were being delivered. Their own fighters such as the Dewoitine 520, the Morane Saulnier 406 and Bloch 152 were quite good and were in production, but largely due to engine power did not have the performance of the Spitfire and Hurricane, or the Bf.109.

I then put on my uniform and reported at Boscombe Down. The Station Commander at that time was Group Captain McEntagart (later, Air Commodore) and not at all a heavy disciplinarian. Neither was Boscombe then a spit and polish Station as Martlesham used to be between wars. The A. & A.E.E. had been transferred from Martlesham Heath, in Suffolk, since it was too near the East Coast. It was considered safer at Boscombe from bombing and the consequent disruption of flight testing. Freeman and Tedder thought that if I started my Service life there, my experience would allow me to judge the nature and seriousness, or otherwise, of any engine or engine installation troubles arising during test flying which might later be reflected in production and operational service.

I should record here that it was Squadron Leader Sorley (the late Air Marshal Sir Ralph Sorley), in 1933, who, in formulating the Air Staff (Operational) Requirement (A.M. Spec.F36/34) for a fighter, decided that it was a vital necessity to score a lethal shot in two seconds. His arithmetic led him to specify eight guns, which armament was adopted by Sydney Camm, Hawker's design chief, when the Hurricane came out as a prototype some three years later. Camm had originally designed for four guns. The same, of course, applied to the Spitfire. In fact, Sorley had earlier tried to expedite

Hurricane and Spitfire acceptance ahead of completion of their flight tests to get them quickly into production, but could not move the Ministry people. Ralph Sorley's foresight at this time cannot be over-praised nor its value to the country overemphasized.

The first aircraft needing my particular attention was the *Hereford* bomber, the Napier *Dagger* engined version of the Handley Page *Hampden* (Bristol engines) and built by Short's at Belfast.

Christmas that year came with one of the worst winters of the century, Salisbury Plain having the lowest recorded temperatures and being swept by blizzards and ice-storms. The Mess lavatories and bathrooms were frozen up. Being a junior officer, I was relegated to a wooden hut with a coke burning stove. That, at least, was warmer than the main Mess building, though there were almost near fatalities from coke fumes. With my firm making up my pay, I could have been comfortable at the hotel in Amesbury but decided that I should learn about Service life and so weathered the winter, finally making it to a bedroom in the main Mess building in the early Spring.

The *Dagger* engines in the *Hereford* were bad starters; one might start and the other engine wouldn't, to be completely the reverse later in the day or the next morning. McEntagart asked me if I could do anything to help, since test flying was falling behind schedule. The *Dagger* had cylinder priming in the upper inlet manifold, with none for the lower banks of cylinders. The upper cylinders could be flooded easily and it was difficult to judge, in the very cold weather, if there was over or under priming. Once flooded, the plugs had to be removed and dried out.

I rang up Napier and told them of the difficulty and suggested that we modify the system to prime both top and bottom cylinder banks. Within a few days they had the necessary extra nipples, unions and pipings shipped to me and Napier mechanics fitted them. I also had the sprays fined down, to give less chance of overpriming. The result was remarkable; both engines could be started without trouble under the very cold conditions and, later, in the warm weather. In fact, the '*Dagger*' became the best starter on the Station. However, I do not know if the new priming system was eventually fitted in production as the *Hereford* was soon phased out, only one squadron being in Service use.

Another trouble concerned the *Dagger* cylinder head cooling which, on the ground, was very limited, the aircraft having to be faced into wind for engine starting and running up. After starting, the engines reached their maximum allowable head temperature very rapidly, so rapidly that the oil was not always sufficiently warmed up for its proper and full circulation. Furthermore, the machine could not be taxied any distance to dispersal because of cylinder heads overheating. All this was to result in a small adventure for me. I was tucked into the miserable little rear gun turret for a test flight and we had only been in the air for about fifteen minutes when I saw smoke trailing from the port engine. Almost immediately, the pilot told me to be prepared to jump, and I didn't even know how to get out and clear. In the event he landed us on one engine. It transpired that, possibly due to the limited warm-up period, the oil circulation was none too good and a failed big end caused a broken connecting rod which went through the crankcase.

My R.A.F.V.R. commission, being in the Administrative and Special Duties Branch, did not entitle me to wear 'wings', though I had been a civilian pilot for some years between wars. I wished to qualify so that I could fly to the squadrons and help with any engineering and engine handling problems they might be experiencing. Not that it necessarily made one a better engineer, but young pilots might listen more attentively to someone who could fly and arrived by air. I tried to get the Station Commander interested when I arrived at Boscombe, but this didn't work. Then, one day, Roderic Hill (Air Vice-Marshal Roderic Hill, Director of Technical Development) came to visit us. Since I had known him when he was Wing Commander, Flying, at the R.A.E., and he always encouraged anyone who had the urge to fly, I asked him if he could help. He said, 'Give me a few days and I will see what I can do'. Two or three days later he 'phoned to say that subject to my passing the full medical for a pilot he had arranged a refresher course for me at C.F.S., Upavon, and I was to report there 'next Sunday'. I was delighted, particularly since I had always wanted to do a C.F.S. Course. 'Next Sunday' was that critical one in May, 1940, when the German armies advanced through the Ardennes into Belgium and France. At Upavon I met Squadron Leader 'Connie' Constantine (later, Air Chief Marshal Sir Hugh Constantine) and Wing Commander George Stainforth of Schneider and World Air Speed Record fame. Everyone was very friendly to a comparative ancient of the First World War and rather surprised that I wanted to be cleared as a pilot, until I told them why. I had their every help, despite the fact that the staff were very busy giving refresher courses to senior flight lieutenants qualifying as instructors and others who had been out of flying for some time. I told them that I thought I could be out of their hair in a week. On the Monday morning, I passed the medical and before lunch drew my parachute. In the afternoon I went up with an instructor in a Miles Master with a Rolls Kestrel engine. After about forty minutes of handling in the air and some landings, he let me go solo.

While doing some spins, I heard a sharp crack and the rate of spin increased. Upon landing, it was found that the fin covering had disappeared. The cause was due to a line of brass brads and the glue both missing a stringer. The leading edge had opened up and the draught did the rest. Another near mishap was my own fault. I was on the downhill approach at Upavon when I thought I was overshooting somewhat, so when an Avro landed across my approach path I opened up to go round again and pulled up what I thought was the undercarriage lever which, in the Master's conveniently combined throttle box, proved to be the flap control. We bounced only slightly.

I had made up my mind to fly each type for at least four hours to become reasonably familiar with them. This I did, with cross-country flights in the Master, Hurricane and Spitfire. Finally, I took dual with a friend of mine and an excellent instructor, Chris Clarkson, on the Bristol Blenheim. This machine was very nice to handle but there had been a number of fatal accidents due to engine failure on take-off, consequently I wanted Chris to give me the handling drill and show me how to act in the case of an engine failure. As a result I was later able to contribute towards a cure, after doing a small experiment while putting in my four or five hours solo. There was, inevitably, a good deal of waiting for take-off at Upavon at that time, with Avro 504s landing in all directions, and it was this idling time, similar to a squadron of Blenheims lining up for an operational take-off, that caused plug fouling. I kept my engines idling at higher speed and a somewhat higher cylinder head temperature than normal. This improved matters considerably and, during a visit to Bristol, I showed how conditions might be bettered. Bristol tested their plugs on a single-cylinder unit at high duty, higher than that normally experienced in actual operational service, with the result that colder running plugs than necessary were specified, which caused oiling up at idling conditions. This trouble was generally cured later, after some squadron tests, by a modified and softer plug.

I had an amusing encounter on one of my Blenheim flights to Bristol when I lunched with Cyril Uwins, their chief test pilot. I said that I could not understand why he had allowed a cockpit layout that reminded me of playing a Wurlitzer organ. Sitting in the pilot's seat, one had to reach backwards with the left hand to change propeller pitch and below the propeller controls were the carburettor 'cut-outs', which could be operated by mistake. Then, working around the cockpit with a gloved right hand, to avoid abrasion, there were the undercarriage and flap operating handles and, as far as one could reach across to the right hand side of the cockpit, the fuel control cocks. Cyril was not amused.

On the Friday of that week, actually after five and a half days at Upavon, I had completed my refresher course, and in the evening went with Chris Clarkson to visit some friends of his at an old Wiltshire manor farmhouse. It was a glorious late Spring evening and I was well content. But the Germans had taken Sedan that week and the real war was on. There was a call for volunteers to go to France as fighter pilots and some instructors and test pilots went, only to be killed or taken prisoner, a useless waste of good material. I returned to Boscombe on Saturday with permission to put up 'wings'.

17 *Working for 'The Beaver'*

On the last Friday in May I was called to the 'phone and a voice said, 'Is that Squadron Leader Banks?' 'The Minister would like to see you tomorrow'. I didn't know which Minister and asked. The reply came back, 'Lord Beaverbrook'. We had heard little of Lord Beaverbrook and the Ministry of Aircraft Production at that time, since the Ministry had only just been formed. I told the speaker that the Station Commander was away and I had no permission to leave. He replied, 'I suggest you come. Everything will be all right'.

I saw Lord Beaverbrook at his London home, Stornoway House in Cleveland Row, on the Saturday morning and we sat and drank champagne. No alternative was offered. He said, 'We are in trouble and Mr Hives (Lord Hives) and Mr Fedden (Sir Roy Fedden) say that you can help me'. Aviation engine production was being 'shadowed' by the automobile manufacturers but there had been no similar planned expansion for engine accessories, 'carbure-eters' (as the Beaver called them), magnetos and sparking plugs. He said, 'I want you to put this right and get them for me. When can you come?' I thought of all the things I ought to clear up at Boscombe and said 'Wednesday, Sir'. 'Wednesday' exclaimed the Beaver, 'the War will be over by then. You come here on Monday. Oh, and I want you out of uniform. I don't like Admirals and Generals around me'. I replied that I was not prepared to do that, particularly as I wished to fly. 'I will give you a high honour if you meet me in this and do the job'. I said that I didn't join for a 'K' and in any case we might not get on and then I would have to re-enlist, if that were even possible. 'Have it your own way' said the Beaver rather shortly.

This meant that I had to 'phone Boscombe to send my effects to the Royal Aero Club, while I went to Harrogate and recruited all those necessary specialists who had been dispersed to that pleasant Spa. This I did and most returned with me on Sunday evening. We opened up shop in the I.C.I. building on Millbank on the Monday morning. I then and there formed a new directorate called the

Directorate of Production and Development of Engine Accessories (D.P.D.E.A.), to fit in with the other sections of the Engine Director-ate (R. & D. and Production).

A number of the Beaver's friends all over the world, political and otherwise, upon hearing of his appointment, offered to come and help him, knowing nothing whatever of the specialist work involved. They generally cabled him offering help and Beaverbrook would cable back, 'Come and help me'. But they never showed their original cable! Many of these 'helpers' merely invoked his name and chased around doing very little other than confusing things. Having had more or less continuous contact with the Air Ministry between wars, I knew that there must be an integrated organization to fit in with the other directorates and being a chaser wasn't good enough. But I became the victim of jealousies since, in forming the Director-ate, I made myself subordinate to certain individuals though I was primarily and directly responsible to the Beaver. However, such irritations had to be borne.

The first vital task was to plan the expansion and, at the same time, dispersal of carburettor, magneto and sparking plug production. At that time, in the middle of 1940, B.T.H. magnetos were being delivered in penny numbers to Rolls-Royce, some in the back of the Chief Engineer's car. This was another instance of lack of planning and forward ordering. For dispersal, we found two mills in Nelson and Colne, on the Yorkshire/Lancashire border, where B.T.H. could disperse. We had to remove S.U. carburettors from the old oil-soaked factory at Adderley Park to a newly built and yet unused Co-op factory in Solihull. There were too few kilns for firing sparking plug insulators and these were doubled up at K.L.G., Lodge and A.C.; Robinson's Paper Works at Bristol were also brought into insulator production.

The shortage of piston rings was acute both for current engine construction and for repairs and maintenance. Some ingenious engineer officers in the Middle East had actually machined rings from cast iron drain-pipes! Yet, when I submitted a paper to the Supply Board (a group chaired by Col. Llewellyn to clear funds for the various directorates) asking for increased piston ring capacity, it was turned down. They seemed quite unable to appreciate the situation or the scale of supply and the pipeline and time factor needed to meet the usage and wastage of war. Immediately I received this turndown I went to the Beaver and told him that it literally must go through. It went through.

George Pate, who had advised the Supply Board against accept-ance of the piston ring paper, was of the family who owned the Albion Motor Company of Glasgow, but for some years had been managing director of Napier. Apparently, there was some dissatisfaction with the production of Short flying boats at Rochester and Beaverbrook had asked Roy Fedden who he could send there. Fedden, for what reason nobody knew, suggested Pate, who had already left Napier and returned to Albion's in Glasgow. Pate saw the Beaver and was about to go to Rochester when Arthur Gouge, the brilliant designer and boss of Short's, threatened to resign if Pate stepped into the works. Instead of apologising to Pate and sending him back to Glasgow, the Beaver did his usual trick and added him, as a Director general, to the M.A.P. Fedden admitted to me some time afterwards that he had made a bad mistake in advising the Beaver on Pate.

Pate was a most obstructive individual and no less was Scott, who sat on the Supply Board. He, Scott, was an earnest and senior civil servant, in fact a Deputy Secretary. Before the War, he had been labelled the 'saboteur' of the Royal Air Force since he opposed most requests for funds for new equipment or expansion. Yet, as I grew to know him better, I liked him. He sat on the Power Jets (R. & D.) Board with me after the War and resolutely opposed the payment to the Treasury of any royalties on patents received on the Company's account, and even advised that the monies be used for improving the Company's consultancy services! Scott had also served in the Royal Navy in the first World War. Hence our mutual attraction I think.

I would not like it thought that I am agin' the Civil Service. Far from it. I always found them very helpful and, after all, I was virtually a civil servant myself, albeit temporary. What I did find in working with them was that they could not, in many cases, after generations of restrictions by the Treasury, adjust themselves to the scale of war expansion. A few who were, however, able to appreciate this such as Arthur Street, Henry Self and Archie Rowlands (our Permanent Secretary) had reached a high level and were outstanding in their work. The British Treasury with its unimaginative and restricted and restrictive outlook has cost the country dear over many generations, being committed to high spending only with the emergency upon us. Some departmental civil servants often do not press their case(s) to the Treasury because they are, after all, Treasury appointees and their job is always to economize. But in war all this was different. I myself once sought a personal interview with the Treasury official alleged to be against granting funds for a project I had submitted – and got what I wanted.

Before we managed to disperse magneto production fully from the Midlands to the North, there came the Coventry air raid. I spent three or four days in the city and got a lot of help from Billy Rootes' (the late Lord Rootes) Committee that had been formed to put Coventry on its feet again. I then returned to London to see Beaverbrook and tell him what I wanted. I told him that a large part of the work force of B.T.H. hadn't returned and might be dead or helping to get their homes in order again. I wanted them back and then I wanted replacement machine tools. The Beaver, who had fallen out with Ernest Bevin, the Minister of Labour, on account of Bevin (rightly) accusing him of robbing other industries of labour for his aircraft factories, said to me 'I want magnetos and carbure-e-ters without labour or machine tools and you are going to get them for me'. I replied, 'Minister, only God can do that'. He said, 'Then fetch Him down and let Him have a try, I'm b——————d if I am going to Bevin for you'!

Beaverbrook had a great fondness for summoning one at all hours, and when you got out of bed at midnight or later and arrived at the Evening Standard offices or motored out to Cherkley, his house at Leatherhead, he might ask you merely a trivial question – trivial because it could have been answered in the morning. A car would collect me at the R.A.F. Club and take me there, where I was ushered into the Beaver's study. After giving him the answers he wanted he told me to go into the living room and, 'Have a drink with the others'. The others seemed to be regulars, like Sir Charles Craven (Chief Executive or Controller of M.A.P.); Archie Rowlands (Sir Archibald Rowlands, Permanent Secretary of M.A.P.); Brendan Bracken (Minister of Information) and various others. I would have a drink and join in the general conversation, then return to London. Sometimes I would have to sit through the Beaver's favourite film, 'Destry Rides Again'. I saw it there three times having seen it when it was first released. The Beaver liked the bar-room scene where Marlene Dietrich and Una Merkel had a catfight on the floor and someone broke it up with a bucket of water. The Beaver really loved that scene. So, I believe, did Churchill.

Later, I was able to convince those on high that Pate should be dispensed with and returned to Glasgow. This was done but we then got another Director General, one C. J. Stewart, who was in many ways worse than Pate. Prior to Farren going to the R.A.E. as its Director, the reputation of the Establishment had deteriorated somewhat in pre-war years. It was often accused by the Aircraft Industry of taking the Industry's ideas and then modifying and patenting them. 'C. J.', then at R.A.E., used to help those who wanted to negotiate a patent and was, inevitably, nicknamed the 'patent agent'.

Stewart was inclined to bully and he also had his toadies, some of whom he brought in as snoopers. At that time a difficult negotiation was going on (through our Purchasing people in Washington) with Bendix to ensure the supply of their injectors for the Packard-Merlin engines as well as spare injectors. Also there was equally arduous work involved in planning the production of the Hobson-R.A.E.

single point injector for Bristol engines. All these matters were made more difficult and protracted by Stewart's interference. Eventually, he was removed and I was then appointed Director General of Engine Production. This change came about because, yet again, there had been no planning and consequently no forward ordering of material. The Rotol company was critically short of propeller forgings consequently production had slowed down. Naturally, they raised a scream to the Minister. But it was a temporary civil servant, Eric Hill, who had warned the Ministry in the first place that there was a propeller crisis – a real shortage. I was given charge of propeller production on account of this but a few weeks after I had formed D.P.D.E.A.

From the inception of D.P.D.E.A. we had been steadily building up accessory production from dispersed and expanded factories. It will have been noted that when taking on accessory production I had included development although it normally functioned within Engine R. & D. But I considered the combination very necessary. Various new equipment required design changes resulting from trouble in operation and the exercise of considerable judgement was needed to avoid unacceptable loss of production in the modification and/or changeover period. But while I managed to hold on to development for about a year or so, when I became Director General of Production and a new C.R.D. (Controller of R. & D.) had been appointed he was prevailed upon to remove accessory development from us.

I recall the dispersal of the Rotax magneto production from their Willesden factory to the Lion film studios at Beaconsfield. In the night bombing following the Battle of Britain, Willesden had suffered badly and we decided that we must evacuate Rotax magneto production. The critical plant was the coil winding machinery which was delicate and generally of the firm's individual design and manufacture. If destroyed, this would bring the whole magneto line to a halt for weeks or even months. So, after two or three consecutive nights of bombing, we arranged for lorries to take out all the magneto production tooling and get them on the roadside in the nearby country while we proceeded to commandeer and clear the Lion studios. These were most suitable since they already had adequate electrical supply and good flooring. But the Lion people proved difficult and protested to the Beaver. However, when I told him the position he was very rough with them and we took over. It transpired that various senior members of the Lion film company and their friends had been using the studios for sleeping and living in during the bombing and were most reluctant to surrender these facilities. Rotax resumed work literally in days and remained there throughout the War.

While there were numberless problems to solve and many demands upon us by the industry, more particularly from the smaller companies who were faced with an expansion many times greater than anything they had known, the biggest problem I faced when becoming Director General was that of assessing and planning forward engine production requirements. Up to that time (1941) the true engine needs were not really known and a projected aircraft build estimate was urgently needed, covering spares requirements for both home and foreign theatres of war, with the total number of engines and spares for the pipeline and current war wastage. It all seemed pretty haphazard and no one appeared to have really got down to the problem until there emerged a man who was to save our collective production skins. He was John Jewkes, a professor at Manchester University, who was called in, by whom I don't know, to form a planning and statistical department. One day he came to my office to introduce himself. I explained our difficulties and prevailed upon him to let me have one of his people permanently in my Directorate. I wanted someone who could indicate the numbers of engines and spares required in a build programme that made sense to the engine manufacturers. The first of these appointees was Brian Davidson, later to join Bristol, followed by Eric Bulley. Each in his way was excellent and relieved us of a great load – 'us' meaning my department and the industry. We could then plan ahead without continual crises arising from unexpected demands and shortages. I was, I think, the first of the Directors General in the Ministry fully to use Jewkes and Devons, who had joined him, and also Alec Cairncross (now Sir Alexander Cairncross). How on earth anyone previously knew what the real demands were and how they could formulate them from the Air Ministry's (A.M.S.O.'s) schedule of requirements I will never know. Eric Bulley proved a great strength to Engine Production and had a most pleasant personality.

The next real production problem came in 1942–43 and concerned the Napier Sabre engine designed by a team under Frank Halford and intended as a high-altitude fighter engine of around 2,000 bhp plus – the most powerful at that time. The Sabre, which had passed a very successful Type Test in 1939, was the engine destined for the Hawker Typhoon, but once in production it gave a lot of trouble and in no way came up to its Type Test expectations. The main trouble was with the sleeve valves, or rather Napier's attempts at manufacturing them. It was a horizontally opposed, 24 cylinder, single (Burt) sleeve valve engine with two crankshafts geared together. Two cylinders fired at the same time, one in each block, mainly for reasons of torsional vibration.[1]

The secret of the success of the Burt and McCollum single sleeve was in its manufacture rather than design, but Napier persisted in trying to make their own sleeves of chrome-molybdenum steel. These had a hardened belt on the inside diameter at the top of the piston stroke and were an almost impossible production job because of distortion or ovalizing, up to .008 in – 010 in, during manufacture. They had also tried silver plating the sleeve at this point. The problem became so acute that we who were responsible for the engine's production had difficulty in meeting the production schedule and could not always get engines through the relatively short (two hour) production test. There were, as a result, large numbers of engineless Typhoons around Gloster's Brockworth airfield. I approached Napier and Halford a number of times, warning that the situation was critical and suggesting they go to Bristol who, after many years of intensive sleeve department, had mastered the manufacturing problems. In desperation, I got Rowbotham, managing director of Bristol Engines, to make me two sleeves for test in the twin cylinder Sabre unit. By luck, the sleeve forgings for the Bristol Taurus engine were suitable for machining to Sabre size. The Napier manufactured sleeves usually lasted for twenty or thirty hours running in the test unit before unacceptable wear caused excessive oil consumption. The Bristol sleeves, which were of nitrided austenitic steel, lasted sixty hours in the unit without trouble; so I asked Aubrey Burke[2] if he would continue the test up to 120 hours. This being successfully completed, we then ordered two complete engine sets for Type Test. Whitehead, the Bristol special manufacturing department genius, got them for us very quickly.

While I had asked Napier to do these tests, Engine R. & D. were very annoyed at what I had done; quite remarkable, considering that I had already offered them this solution. Then Rowbotham objected to Bristol giving their manufacturing secrets to Napier – for a key aircraft in the middle of a War!

The final showdown came early in 1943 when English Electric took over Napier from the appointees of Lazards, the banking concern. At that time, when German 'tip-and-run' raiders were shooting up the South Coast towns, Churchill had ordered standing patrols of Typhoons to intercept them. The Typhoon was then the fastest low level fighter, though originally intended for high altitude combat. But with the unreliability of the engine and the loss of some aircraft in the Channel due to this rather than enemy action, coupled with the fact that Leigh Mallory, C.-in-C. Fighter Command, had insufficient aircraft to form more O.T.U.s, he had to report to Sir Archibald Sinclair (Secretary of State for Air) that he was unable to carry out the Prime Minister's directive. So the message came down from Sinclair to Portal (C.A.S.) and then to our Chief Executive, Sir Wilfrid Freeman, who sent for me and said, 'I am making you responsible for the development of the Sabre. Go and take over from Engine R. & D.' I replied, 'I can't do that, Sir. You must first tell them to hand it over to me'. Freeman demanded 'You do as you're told.' I replied 'No, Sir, I cannot and it will only cause unnecessary friction'. Finally, Freeman accepted the position and I had the Sabre in my lap.

Having already negotiated the preliminaries for Bristol manufactured sleeves, I gave Bristol a contract covering two engine sets for Type Test which, to cut a long story short, the Sabre passed without trouble. There were other weak points in the engine that had to be attended to, but in about six months the Sabre became one of the most reliable engines in the Service.

After clearing sleeve manufacture, Bristol fashion, I rang Biddlecombe of the British Purchasing Mission in Washington D.C., and ordered the Sundstrand machine tools that Bristol had advised. These were put on board the *Queen Mary*, then ferrying thousands of American troops over the Atlantic, and the machines were working in Napier's Liverpool factory within a month.

When it came to the Tempest V with the increased rating Sabre IIB, Camm of Hawkers was of a mind to fit another engine and Sorley asked me to go and persuade him otherwise. Camm had already concentrated upon the Tempest II fitted with the Bristol Centaurus engine, but I told him that the order would in any case be split and if he didn't want the Sabre Hawkers would only get about half the order. In the event Sydney capitulated and it was the Tempest V that was so successful in Normandy and in dealing with V1 flying bombs. The Centaurus in the Tempest II suffered considerable introductory trouble and I was able to help Philip Lucas, Hawker's chief test pilot, to overcome Bristol's inertia in getting their engine right. This took some time. The Tempest V's performance was a very great improvement on that of the Typhoon. It had a thinner wing of elliptical or Spitfire shape but with squared off tips, and its handling was apparently excellent. It had, however, leading edge radiators, which the Service distrusted for the ground attack rôle on account of vulnerability so the production version reverted to the ducted radiator under the nose.

I assumed personal responsibility for Sabre development and production at the beginning of 1943, at the same time as the takeover of Napier by English Electric. So I was somewhat surprised to see that a young man of 26 years had been appointed managing director. He, George Nelson, was the son of the English Electric boss. This worried me, considering the poor management suffered by Napier over some years, so I called upon our Ministry of Labour office for his file and was surprised to have it refused me. When I enquired why the M. of L. man said that Sir Wilfrid Freeman had given the order. So I went to see Wilfrid and he smiled mischievously at me and asked why I wanted to see it. I told him that we had sufficient to put up with at Napier and I didn't want, in this fresh start, continued management difficulty. He then had the folder released to me. I should say that George was highly qualified, with technical degrees from Cambridge and Zurich, and I was satisfied that we had someone of intelligence in charge.

That same week I had my first full Napier meeting at Acton, in my authoritative rôle, and was somewhat surprised when George Nelson wasn't there. He had left a message that he might be back before we finished. I went through every detail of the Sabre's troubles, and there were three important modifications that had to be included before further deliveries were made. We were still in production on the old Napier sleeve valves which in any case limited engine deliveries to such sleeves as were passable, though the Bristol sleeves were coming along rapidly.

The A.I.D. man, who had been bypassed on several previous occasions, asked me when engine deliveries were to be stopped for the modified parts to be included. I told him, 'Immediately, and you may leave the meeting and stop any engines from here and Liverpool.' Standring, the Works manager, protested but was over-ruled. I had to tell him that we could not tolerate any further deliveries of unmodified engines. Having agreed the modifications and ironed out the sleeve production programme I broke up the meeting to return to the Ministry. I had got as far as the Works gate when the commissionaire stopped me and said that Mr Nelson, Senior, was on the 'phone and wished to speak to me. The conversation went something like this: 'Is that Group Captain Banks?' 'Yes, Mr Nelson'. 'Oh, Banks, I hear you have stopped Sabre deliveries. I can't have that'. 'Well, Mr Nelson, those are my instructions and they must stand until the modifications are completed.' 'I will have to see Sir Wilfrid Freeman'. 'Certainly, Mr Nelson.' I then found out where he and George had been during the afternoon when he said, 'I have just seen Air Marshal Leigh Mallory and promised him the engines he wants for his operations and the O.T.U.s.' I said that I was sorry he had done that without finding out the true position. In fact, he had promised Leigh-Mallory more engines than we had materials for or that the programme demanded.

I went back to the Ministry and immediately saw Sir Wilfrid and told him to expect an infuriated industrialist on his doorstep. Indeed, while we were talking, Sir Wilfrid's P.A. came in and said that Mr Nelson was asking for an appointment, 'tomorrow'. Freeman said to me, 'There you are! We have just persuaded English Electric to take on Napier and you have messed it up'! I said, 'Very well, Sir, hand development back to Engine R. & D.' He said, 'Don't be stupid. Tell me all about it and brief me for the meeting to-morrow.'

The Nelsons came the next day and after they had gone Freeman sent for me. He referred to some notes he had made and said that he had been told that the modifications could be made 'as and when' the engines were available but it wasn't necessary to stop deliveries. I replied that I knew the nature of the modifications he had in front of him, and the defects had caused the loss of machines and, we believed, two pilots.

Sir Wilfrid finally said, 'What do you want me to do?' I replied, 'Please ask C.A.S. (Portal) to tell Leigh Mallory not to listen to enthusiastic industrialists and we (M.A.P.) will give him the aircraft delivery rates every week as engine modifications are cleared.' And so another crisis was under control. But not before the Acton A.I.D. man rang me one day to say that Standring had sent six unmodified engines to Gloster and what was he to do? I replied, 'Transfer me to Standring'. I then told Standring that if the engines weren't back by the morning I would see that the A.I.D. took Napier's 'approval' away. In the event, the engines were back at Acton that night. I doubt, in fact, that we could have removed their approval since there weren't sufficient A.I.D. inspectors to take over!

One final difficulty with the Sabre came immediately after the Normandy landings in 1944. The meticulous preparations for the landings missed one important factor, Normandy soil and its effect on the sleeve valve engine. This had a high proportion of gritty dust which was an excellent silicon abrasive said to be used for making commercial sandpaper. After a firm foothold had been established and a landing strip bulldozed for the fighters, when they ran up their engines they couldn't be seen for clouds of dust. This was a deadly condition for a sleeve valve engine and the Sabres in the Typhoons were being pulled out after only a few hours' operation. An inertia filter had in fact already been tested and approved, but nobody bothered to have it fitted. However, there was no question of any further delay, so we went to see Mr Vokes the filter specialist. He responded as he always did, with enthusiasm and energy, his workers turned out all the filters we needed in, literally, days, and we were in the clear.

I was flown over sometime later in an Avro Anson and landed on an airfield to visit the Typhoons after they had broken the German tank formations holding up the American forces at the Falaise Gap. This particular operation alone possibly justified the whole Typhoon/Sabre programme! These machines caused havoc among the Tiger tanks and saved the day. I saw a whole tank turret with its gun yards away from the tank, seemingly from a rocket hit at the turret base, though it has been said that rockets weren't all that accurate.

But an awful lot were fired. They were clearing the carnage of this action on the day I arrived, and one could smell death and putrefaction when still some hundreds of feet in the air during the landing. The weather was warm and they were using bulldozers to bury the remains in pits.

One day, my Acton R.T.O. (Resident Technical Officer) came to see me and said that his colleague in the English Electric Halifax production plant at Preston had told him that Mr Nelson Snr. had addressed his people on the subject of industrial courage and cited the case of the Sabre and how, due to the trouble it had given, *he* had had to hold up deliveries in the middle of the War! What, asked my R.T.O., was I going to do about it. I said, 'Nothing'. It might be thought from the foregoing that I did not care for Mr Nelson Snr. (later Lord Nelson), but this is not so. In fact I liked him and had always admired the way he had created the English Electric organization after the First World War–from Coventry Ordnance, Dick Kerr's of Preston, Willans and Robinson and others. Like many other industrialists of his generation he was always 'the Boss'.

Throughout this critical period I was greatly helped by W/Cdr. Harry Mundy, now Chief Engineer of Jaguar. He was apprenticed to Alvis and then went to E.R.A. He joined the R.A.F.V.R. in 1939 as a pilot, then was transferred to engineering. Over the years, Harry Mundy has established himself as a first class design engineer.

The fixed pitch wooden propeller, having to cater for relatively large differences in indicated air speeds between full throttle level speed at rated altitude and that at sea level, posed serious shortcomings in the handling of fighter aircraft. First, since the blades were designed to accept maximum engine power at rated altitude, they could only absorb some 60% of take-off power. This meant that both the take-off run and time to height took a great deal longer than if the full power of the engine could have been used regardless of forward indicated speed (because maximum permissible engine rpm could not be developed).

The obvious solution, therefore, was an infinitely variable (constant speed) propeller. But the work originally done in the US and Europe on variable pitch propellers was at first limited to the two-position blade; and though of considerable help, it had little, if any, effect on rate of climb because, once airborne, it was necessary to go into coarse pitch to prevent the engine overspeeding.

de Havilland, being the first to produce a two-pitch propeller in the U.K. (under their Hamilton-Standard licence) were closely followed by Rotol, the propeller company formed by Bristol and Rolls-Royce, mainly at the instigation of Roy Fedden, and largely based on the original Hele-Shaw Beacham patents. The Hurricane and Spitfire were each flown with both these two-pitch propellers early in 1939, and the enormous advantage over the fixed pitch wooden propeller was apparent. There was, however, a certain apathy in official circles to the early introduction of the two-position, variable pitch, propeller, mainly, it is thought, because these two fighters were originally designed as interceptors, at a time when dog-fighting in the coming war had hardly been conceived. But when the war began to hot up, after the fall of France and when German bombers could then reach the U.K. with fighter cover, there was a universal howl from the fighter squadrons for the constant speed propeller. As a result, when the Battle of Britain started, all Hurricanes were fitted with the deH-Hamilton constant speed propeller and the Spitfires with the equivalent Rotol, there being more Hurricanes than Spitfires at that time. Had there been no constant speed propellers, and incidentally no 100 octane fuel, the result of the Battle could well have been different, since all German fighters had C/S propellers from the start. Therefore, the pressuring of the Ministry by such individuals as Parkes, Lucas and Quill, and of course the squadrons, largely contributed to saving the day.

The deH-Hamilton propeller had two quite serious defects, the first being oil leakage, which obscured the windscreen, virtually blinding the pilot, the second being its slowness of pitch change due to the hydraulic system not being sufficiently powerful to overcome the twisting moment of the blade with the shifts in the centre of pressure on the blade that accompanied changes in angle of incidence. For instance, if the throttle was suddenly closed and the speed reduced by hauling up the nose, the propeller would instantly go into fine pitch, but when the throttle was opened and the engine speed increased the propeller would stick in the fine pitch. The pilot would then have no alternative but to break off the engagement – if he could! He would then slow the aircraft to near the stall before the blades would come off the fine pitch stop. Since neither of these defects could be overcome without major design changes, accompanied by further development (more than 3,000 miles away at Hamilton), it was officially decided eventually to fit all fighters with the Rotol propeller (it did not suffer oil leakage and had a very rapid pitch change response) when Rolls produced the Merlin III which incorporated a standardised propeller shaft capable of taking either the Rotol or the deH-Hamilton propeller. Since the deH-Hamilton problems did not affect four engine bomber aircraft, this propeller was chosen as the 'bomber propeller'.

Much of the success of subsequent events was due to the efforts of John Parkes, at the time General Manager of the Engine and Propeller Divisions of the de Havilland Aircraft Company. He had tried to influence a change in official policy but was actually told that the introduction of constant speed propellers on Spitfire and Hurricane aircraft must be delayed until an alternative product became available (the Rotol propeller was not yet in production). I remember thinking it quite extraordinary at the time. Rather like the situation in the First World War when the parachute became available which could have saved hundreds of lives, but the Air Ministry (Air Staff) wouldn't allow its use because it might not be reliable! And some years afterwards, in 1928, when a paper on the Hele-Shaw-Beacham variable pitch propeller was given at the Royal Aeronautical Society, the short-sighted D.T.D. (Directorate of Technical Development) came down against the propeller as too complicated. Wing loadings at the time were little higher than 9 lbs/sq ft for civil aircraft although some fighters topped 14 lbs.

I have attempted here to give some idea of The Beaver's character:

Beaverbrook was a sort of political opportunist although he didn't appear to have any really great political pull so far as parliamentary influence in the Government was concerned. He was, of course, close to all the important people and politicians, like Churchill, whose attention and influence he was able to command and so become one of his Minister's. As minister of M.A.P., he used a number of good, bad and indifferent people as helpers and chasers who used to invoke his name to try and get things done – so much so that their influence tailed off because those running plants got tired of lesser people continually using the Beaver's name without really knowing enough of the job they were supposed to be 'encouraging'. Invoking Beaverbrook's name was a favourite trick of the individual chasers he surrounded himself with, though few knew how a ministry was run! The Beaver did not, as was sometimes thought, override senior civil servants, like Archie Rowlands, our Permanent Secretary, and he did take notice of what they and also specialists like myself told him, provided we were firm in our advice based upon knowledge and experience. I found that, provided I stuck to my guns, he did not continue any of the bullying that he often tried at the start. His stunts, like asking housewives for their aluminium saucepans, were not helpful, and he was pleaded with by a number of us, and by those in the aluminium industry, not to go on with it. Publicitywise, however, it brought the public into the act and made them feel that they were personally helping the war effort.

To sum up: I think he did energize the Ministry staffs, particularly those who had joined him from other ministries and who had been for some years on routine work and could not easily overcome a degree of inertia and apprehension in regard to decision-making, particularly when they had to fight the Treasury. In this latter respect Beaverbrook did a great service in getting the Contracts procedure cleaned up so that firms could go ahead with the minimum of delay and paperwork – almost by word of mouth and a confirmatory letter. The Ministry even suggested that the banks might be induced to change their policy so as to provide the industry quickly with the working capital it needed. This was called the 'Scheme C', and was arranged after talks with the Bank of England, the Treasury, the three War Departments and the clearing banks who were ready to meet the requirements of Government contractors and sub-contractors (as Postan[3] has pointed out). Working capital and progress payments could thus become quickly available to avoid financing difficulties which would cause hold-ups in production flow.

Beaverbrook's other great service, in my opinion, was when he got the repair facilities further removed from R.A.F. control and distributed more widely among the industry. His policy was to make M.A.P. responsible not only for the Civilian Repair Organization, but also for repair work carried out by '43 Group' of the R.A.F. He also brought the industrial capacity available for repairs to the same degree of dispersal as the rest of the aircraft industry, resulting in the considerable enlargement of the circle of repair contractors. For instance, Rolls-Royce's spares problem was solved by setting up the Sawley Spares Depot, operated by the firm as the sole supply and distributing centre for all R-R spares.

I think, therefore, that everyone was woken up to the need for a major change that had to be, i.e., a vast change on a rapidly increasing scale beyond the between-wars system; and Beaverbrook was in office long enough to get things started and moving. He didn't steepen the previously planned production curve, but he put an upward 'kink' in it by robbing the stores at a critical time, giving some breathing space for steady production with minimum interruption later and pressing for more rapid production to relieve tight spots, thereby creating a sense of urgency and enthusiasm that had previously been absent particularly after Dunkirk. He could not have remained effective after his time, or chaos and upsets would probably have resulted.

During my early professional life and in the last War and since, I have enjoyed the friendship of Jimmy Doolittle (Lieut. General James Doolittle, U.S.A.F.). It would be presumptuous and superfluous to relate his life's story here, and I cannot remember when I first met him, but think it was at an early S.A.E. Summer Meeting when Sam Heron introduced us. Our subsequent meetings have always been interesting and profitable – to me. I first saw Jimmy fly at one of the annual Cleveland Air Races when he piloted a very dangerous-looking racing aircraft, the Granville Gee Bee and won the most important race of the day, going on to break the World Landplane Airspeed Record after the race, even before landing.

Meeting him some time after his Gee Bee piloting feat, he said it was the nearest thing to 'Balancing a ball on a pyramid'.

I saw Jimmy a number of times in World War II when he was commanding the U.S. Eighth Air Force. In fact, I nearly got into serious trouble on his account, over the matter of sparking plugs! When the B17s came over here their engines were fitted with plugs having mica insulators which, after their earlier and shorter bombing operations over the Rhineland, needed removing, cleaning and sometimes replacing before the next operation. All this caused considerable operational delays. Asked to help, I fixed with Jimmy's people that if we could get sufficient 18 mm plug bodies or shells (ours were 14 mm) we would make up some plugs for test with sintered aluminium oxide insulators. This was done and, after satisfactory proving trials, some thousands were supplied until U.S. production took over. All this was publicly acknowledged later by President Roosevelt. The sequel was that some of our top brass accused me of denuding the R.A.F. of plug supplies. Actually, our capacity had built up so well that the R.A.F. did not go short of a plug.

Jimmy Doolittle had retired from the U.S. Army Air Corps in the 'twenties or early 'thirties to join Shell as their aviation specialist, later going on the Board. My contact with Shell Aviation in the U.K. was Ernest Bass (E. L. Bass). Ernest was a good servant of Shell and a capable engine and fuel specialist. Pre-war we went on many trips together to the airlines, with visits and meetings at the Shell Thornton Fuel Research Laboratory near Chester. These were presided over by (the late) Dr C. G. Williams. Bass, Williams and another Shell technologist, Terry Rendel, all did a first class job for us at M.A.P. throughout the War; also representing M.A.P. in the States and working with our own fuel and oil specialists, Drs Drinkwater and Ridler.

Esso (now Exxon in the U.S.) were, of course, helping us in the same way. After hostilities, Esso made a very valuable contribution to lubrication technology for gas turbine engines by their development of EEL3 (called 'EEL-Juice'!) This was an oil capable of remaining fluid at very low temperatures, down to minus 40°C, with the capability of lubricating highly loaded gear teeth, such as propellor turbine reduction gears and high speed accessory gear drives. This research was initiated by the late Harry West when, as a member of the C.I.O.S. team in Germany after the War, he came upon the work by Zorn with '2 Ethyl Hexyl Sebacate'. Ronald Moreton, working on this at the Abingdon laboratory eventually came up with EEL3. Among those who supported West and Moreton were Peter Windebank, Ken Hunt and John Brown.

*   *   *

Early in 1944 I was sent for by Freeman who said that he and Air Marshal Sir Ralph Sorley (the new Controller of Research and Development appointed after Linnell's death in a car accident), wanted me to take over Engine R. & D., which would also include gas turbines. Power Jets was to be made a nationalized Company by the Government as Power Jets (R. & D.) Ltd., and Roxbee Cox (now Lord Kings Norton) appointed Chairman and Managing Director. I gladly accepted the offer (command!), since my background was more in engine R. & D. than as a production specialist. In any case we had reached our engine production peak. So, upon becoming Director of Engine R. & D., I was promoted from Group Captain to Air Commodore. It always seemed peculiar to me that one held a lower rank as a Director General of Production than as a Director of Engine R. & D. The explanation was, apparently, that production was only a temporary or wartime set-up whereas Director of Engine R. & D. was a permanent appointment and as such regarded as a career job for a technically qualified permanent civil servant, or officer. Major G. P. Bulman had held the post prior to me for some seventeen years.

Freeman told me, when he broached the subject of the new appointment that Sorley and he wished me to take over both piston engine and gas turbine R. & D. responsibilities in order to control and co-ordinate the effect on each and get the then new class of engines established 'in the body of the Kirk' – Freeman's term for the legitimate aero engine manufacturers. From Roxbee's old D.S.P. (Directorate of Special Projects) I inherited George Watt, a very able Group Captain from New Zealand, and also some gas turbine performance specialists, particularly valuable 'long haired' people from the R.A.E. who had been working on gas turbine performance and were assisting the engine manufacturers with the intricacies of its aero- and thermo-dynamics.

One of the first problems I had to tackle concerned the contracts for some very large piston engines which Engine R. & D. had previously given to Bristol and Rolls-Royce. These engines, luckily only then in the design stage, were respectively 24 and 28 cylinder jobs, eventually to be rated at between 4,000 and 5,000 bhp. It was quite obvious, even in 1944, that the gas turbine was going to take over as *the* fighter aircraft engine and that these enormous piston engines would involve some seven years or more development and then be too small in power, too heavy and too late. It was not difficult to persuade Rolls that such engines were not on but when I went to Bristol I met with some resistance both from Norman Rowbotham, the managing director, and Sir Stanley White whom he had taken me to see. It so happened that Bristol were also designing a high economy propeller turbine of about 2,200 eshp, the Theseus, and Owner, the chief designer, could not get sufficient detailing draughtsmen to finish the job and bring it to the metal cutting stage. My proposition was to cancel the large piston engine which Roger Ninnes was designing as this would automatically release draughtsmen who would either have to face the possibility of direction elsewhere or have to work on the Theseus gas turbine. When I broached the cancellation to Sir Stanley White he said that there was still considerable life left in the piston engine. I replied that we only wanted intensive development of the Centaurus (still in trouble) and the Hercules, but a big new piston engine would put back the gas turbine by some ten years and leave Bristol trailing. He then accepted the situation.

About a year later, in 1945, I was to order the Proteus propeller gas turbine, after viewing the general design and arrangement drawings laid out on the grand piano in the Bristol guest house. Originally intended for the Brabazon, it was eventually fitted in the Britannia.

During the period of 1944 onwards, Rolls-Royce had Hooker and Lombard, at Barnoldswick, designing and building centrifugal blower type gas turbines. These were to become the Derwent and the Nene. Earlier, there had been an attempt by the Ministry to get Power Jets working with Rover since that company was to redesign the Whittle jet for production and then go ahead and manufacture it. But there was much friction between the two concerns and Rolls-Royce were asked to take over, Rover eventually being given the (unsupercharged) Merlin tank engine, the Meteor, to produce. Adrian Lombard, who was with Rover and a natural born designer, left and went to Rolls. He came to Derby from Barnoldswick when Hives decided that, after the war, all engineering was to be re-established at Derby. Lombard became Technical Director of Rolls aero gas turbine engines after Hooker left for Bristol. He died at a relatively early age in 1967 and his demise was a great loss to Rolls. Hooker (now Sir Stanley Hooker) returned from partial retirement in 1970 to go on the main Board of Rolls Royce as Technical Director when the '211' crisis came.

*   *   *

I have not so far spoken of the bombing because, after all, millions of citizens were involved and had their stories. I can only say that when in 1940 the London bombing started we were so busy under Beaverbrook and working all hours of the day as to take little notice of it. At first, with daylight bombing in the Battle of Britain period, at every alarm we were all hustled by the wardens from our offices to the shelters in the basement of the I.C.I. building but this lost so much time that it was discontinued. Some of us who were too tired or too late to go back to our Clubs or homes dossed down in the basement at night, but I snored rather badly at that time and kept everyone awake so I had a camp bed put up in my office. I used to walk back to the Royal Air Force Club in the early morning for a bath and breakfast. A number of us decided that if we had, say, a dinner appointment, we would keep it whatever the bombing, otherwise life wouldn't be worth living. Two incidents happened to me: the first concerned one bomb of the stick that fell across Piccadilly at the Ritz and Devonshire House and Green Park Tube Station. This blew our taxi on to the pavement but without injury to us and minimum damage to the vehicle. One of the Beaver's helpers, 'Troubles' Elliott, was with me. The only injury I sustained was a bad hernia from helping to lift an overturned car off a woman. In the V1 flying bomb period I had my bedroom ceiling come down on me one night in the R.A.F. Club when a bomb hit the wall surrounding Buckingham Palace. I found people rather more jittery in the mini-blitz of 1943 than in 1940. By then the pressure of work had dropped or got more routine and there had been no bombing for some time.

I was bombed earlier in the Hawker Siddeley (Armstrong-Siddeley) guest house in Kenilworth one night by a machine unloading its bombs in a Birmingham raid. But beyond fracturing the water main, there was little damage and H. K. Jones, the Managing Director, and I spent about an hour talking and drinking whisky before retiring again.

A few days before the actual start of the V1 operations, the Germans sent one over as a sighting shot and Air Intelligence managed to pick up the pieces and collect some fuel from the tank, which my friend Michael Golovine of Air Intelligence rushed to me for analysis. We found it to be ordinary aviation gasoline and nothing at all exotic.

The most frightening time to me was in the dive bombing of Manston, during the Battle of Britain period, where I had gone to see a Bristol Blenheim after an engine cut on take-off. I wanted to see the state of the plugs before anything was done. The aircraft was, however, somewhat damaged and punctured in the bombing.

*   *   *

I have mentioned that those of us on the production side in M.A.P. had some difficulty in understanding and interpreting the Air Ministry's (A.M.S.O.'s) spares requirements. For instance, in 1940, a shortage of Cheetah magnetos was reported for the Avro Anson; a nil return for these was recorded from A.M. Stores. We who worked under the 'Beaver' had already organized search or working parties to raid the Stores. Storekeepers often seem to consider their job is done when the bins are full, but we had to deplete the bins in those critical days of 1940 to ensure no hold-up in aircraft and engine production, to clear A.O.G.s and to relieve line shortages in the factories. The alleged shortage of Cheetah magnetos was not in fact a shortage at all and in our searches we uncovered a further 250.

Then, one day in 1943, we received a requisition from the Air Ministry for so many spare Merlin crankshafts that a new factory would have been needed to supply them. The need for spare crankshafts for the Merlin engine was minimal, so we could only conclude that, in their curious method of assessment, the Air Ministry had taken a wrong turn. But we couldn't convince them that it was unreal and A.M.S.O. sent his deputy, an Air Marshal, to see the Minister, Stafford Cripps, who asked Ben Smith (the Parliamentary Secretary) to chair a meeting and sort things out. We had two meetings and Ben seemed overawed by the presence of the Air Marshal, who incidentally had no real facts to offer and so could not justify the requirement. After the last meeting, with no conclusion reached and when I had not given way nor had any intention of doing so, Ben asked me to stay behind. When we were alone, he produced a bottle of whisky and then proceeded to tell me the facts of life; how he went to sea before the mast and had become a boxer, and how I must accept the rough (decisions) with the smooth and do what the Air Ministry wanted! I told him that I wasn't going to accept an obviously stupid demand, that I too had been to sea before the mast and roughed it; and I wasn't going to waste precious labour and resources. Ben was a very nice man. He had been a taxi-driver and then, I believe, became the secretary of their Union. In the end, we disagreed but the Air Ministry didn't press their case and there was no crankshaft shortage.[4]

I met and collaborated with a number of unconventional regular or career officers at various times during the War. One such was Air Commodore Graham Dawson (the late Air Vice-Marshal), the Senior Technical Staff Officer (S.T.S.O.) at Fighter Command. Dawson was directly responsible to Dowding and was very energetic and most impatient of delays, particularly those due to procedure and working to the book. The normal procedure in notifying a defect or engine failure in the Field was for the Command or the Station engineer officer (C.T.O.) to complete a 1022 form describing the trouble and sending it, through Group, to the Air Ministry where it would eventually get to the proper department or section. But some time could elapse before any action was taken. In the Battle of Britain period, we were particularly concerned with the sparking plug problem. Before the War, when I brought back from Germany the new Siemens sparking plug with the sintered aluminium oxide insulator,[5] the universal insulator material was mica, a material that was becoming unreliable with increasing engine duty (temperature) and the effects of the products of combustion of leaded fuel. Sintered aluminium oxide, known by the Siemens trade name, 'Sinterkorund', was highly resistant to temperature or the effects of leaded fuel and offered great advantages for future plug and engine development.

The plug manufacturers, learning of sintered aluminium oxide, quickly adopted it. K.L.G. acquired the Bosch licence, which was a modification of Sinterkorund; A.C. (U.S.) bought the Siemens licence and Lodge developed a formula of their own which they named 'Sintox'. The principal difficulty experienced in the design and manufacture of the plugs was the effective sealing of the centre electrode in the insulator to prevent gas leakage and early development brought a string of troubles associated with this.

The problem was to judge, if a modification proved effective, when to bring a new plug into production and when to phase out the previous model without losing too much production in the process. So as to attend to faults with all speed, I agreed with Dawson to let the technicians of the sparking plug companies go directly to the R.A.F. station concerned, identify the trouble and get on with the modification ahead of form 1022 action; but not to upset normal 1022 procedure unnecessarily by such direct action, which would cease as soon as the emergency was over.

With the introduction of the sintered aluminium oxide sparking plug insulator came the fine wire electrode of platinum/iridium or platinum/rhodium alloy. This gave good results and reduced the rate of point erosion, which was still further improved by a 1,000 ohm carbon resistor at the upper or terminal end of the plug. Both these developments were largely the result of collaboration between K.L.G. and Napier. The high value of platinum made it necessary to recover and re-refine it from the many thousands of time-expired or defective plugs. All this was done for us by Johnson Matthey, the precious metal refining specialists.

Reverting to the preparation of production programmes: I remember how complex these would become with a policy change. For instance, when it was decided to replace the Manchester bomber (two Rolls-Royce Vultures) with the much superior Lancaster powered by four Merlins, the decision was also taken to introduce a Hercules version of the Lancaster, the Mark II, to counter any possible interruption in Merlin supply. All these decisions needed very detailed arithmetic between ourselves and the engine and aircraft manufacturers. In the event there was no interruption in Merlin production and U.K. manufacture was supplemented (mainly for the Canadian built Lancaster) by the U.S. Packard Merlin 28. A contract for the Bristol Hercules Lancaster II was given to Armstrong-Whitworth, the first being delivered in October 1942. There were some 300 Lancaster IIs built, but it was phased out in the Autumn of 1944. Both the Lancaster I and III, with British and American built Merlins respectively, were superior in performance to the Lancaster II.[6] On the other hand the Halifax I, originally fitted with Merlin engines, was a poor performer mainly because the engine installation had a high propeller thrust line, above the wing leading edge. The fitting of the Hercules engine, with its concentric reduction gear put the thrust line level with the leading edge, considerably reducing drag and improving overall aircraft performance.

As Director General I had under me a number of excellent people; some had production experience, some were rank amateurs, and all were wartime recruits to the civil service. My senior director was Pat Welman, brother of the Welman who was C.O. of the CMB Base at Dover in the First World War. The deputy director in charge of engine accessories was Davenport. Both men had production experience in engineering plants before coming to the M.A.P. The other deputy and assistant directors, Parsons and O'Dea were in charge of various accessories such as carburettors, magnetos and sparking plugs. My director of propeller production was Sir Patrick Hamilton Bt. Then there was Sir William Dunning Bt and a girl called Maria Stormonth. All were intelligent and energetic and could quickly understand, and often anticipate, what was wanted then sort out the difficulties with other departments and see things through to a conclusion. We acquired Maria Stormonth from a group headed by a Wing Commander whose job it was to concentrate on quickly clearing the needs of any grounded aircraft (A.O.G.s). With the larger Ministry organization then working fairly smoothly, this group was being disbanded and we were asked if we wished to recruit anyone available. Maria seemed alive and energetic so we took her on. We were eventually able to create a new grade and make her a Junior Production Officer to give her proper standing and pay. She had been an actress, having a very good singing voice. She was also doing her stint during the bombing, helping to rescue people and looking after those in pain and trouble. She did all this mostly at nights, working long hours at her job in the Ministry during the day. Maria is now my sister-in-law.

An episode of the time is worth relating. In the Autumn of 1940 the far-sighted Hives, works director of Rolls, came to see the 'Beaver' and said that he felt, in anticipation of 'the next shooting season' (meaning the start of intensive air fighting in the coming Spring) the Spitfire would need more power. Rolls could provide a Merlin of greater performance sufficiently quickly to complete two squadrons for a start; the re-engining could be done at the Rolls flight test department at Hucknall under Ray Dorey. The Beaver was delighted and gave his O.K.

A few months later, when the first of the engines was due,

Sidgreaves, managing director of Rolls, came to see the Minister and said that there was a delay due to the lack of carburettors, 'which were an Embodiment Loan supply'. Embodiment Loan meant government supplied or furnished equipment. When Hives first put his proposition to the Beaver he hadn't realized that the Merlin 45 engine he was proposing required a larger carburettor, the one eventually to be used by Ford for their Merlin 20 intended for the Lancaster but not yet in production. Further, it was *not* Embodiment Loan and, to a certain extent, I sympathized with them because they had little or no control of the production of the particular equipment. But there had to be such a scheme otherwise they, the major manufacturers, would dominate the scene. The Minister sent for me and ranted, in front of Sidgreaves, that he had given me the responsibility of getting 'carbure-eters' and I had fallen down on the job. After I told him that it was a Rolls responsibility, he turned to Sidgreaves and said, 'What do you say to that?' 'Sg.' replied that I was mistaken, so the Beaver told him to ring Hives. Hives came on the 'phone and said that, so far as he knew, the (S.U.) carburettor *was* Embodiment Loan. I said, 'Let me speak to him'. I asked 'Hs' to speak to 'Sft' (Swift, the works manager) who confirmed what I already knew. The Beaver turned on Sidgreaves and told him to be sure of his facts before coming up and wasting everyone's time! Poor Sidgreaves got very red in the face. Nevertheless, we had to help Rolls out of their difficulty and it fell upon myself and Swift to arrange a joint production programme between Rolls and S.U. in Solihull and for some carburettor bodies to be sent to Derby for machining.

S.U., part of the Nuffield Organization, were not particularly good on production though technically very capable. I once found my friend Carl Skinner (managing director and with his brother the original founder of the S.U. Company) setting up a carburettor body on a multi-spindle drilling machine when they had just got into the Solihull factory from Adderley Park. I said, 'What the hell are you doing, Carl?' He replied, 'Getting on with the job'. I said, 'Carl, managing directors are there to organize and not interfere with production'. Eventually, because of their slow progress and production delays, I asked Miles Thomas, (now Lord Remenham), Vice-Chairman of Nuffield, if he would come and see me. He brought along Lord Nuffield who asked what they could do. I replied that I thought it would be helpful if he would loan S.U. Ernest Ruffle (managing director of Morris Commercial Motors) and Peach (planning engineer of Wolseley) to assist and reorganize things at Solihull. He acted quickly and Ruffle and Peach were at S.U. in a day or two. They made a thoroughly good job of getting production rolling.

We proved in the war that, given long production runs, the U.K. was as capable as any other country of efficient production. But one would not dispute the fact of American production efficiency, where she supplied all of us, all the allies, with armaments, munitions and ground and air transport. Even that only absorbed some 70 per cent of her total industrial potential. Nevertheless, Rolls and Ford of U.K. produced Merlin engines at a cost of £1 per bhp, and £1 per lb weight, comparing favourably with American production costs which were somewhat higher due to their higher wage rates.

It might be queried why, for instance, the more complex and costly Sabre was not eventually discarded. But that is being wise after the event and the engines we had were the result of a steady climb up the learning curve of design and development in the course of creating a state of the art. One could not be sure that any one engine type, during the period between wars, would predominate. Though the Merlin was installed in outstanding aircraft during the war the air-cooled engine had been generally favoured between wars. The liquid-cooled engine was more sensitive or vulnerable to battle damage than the air-cooled type which was therefore more favoured in the ground attack rôle. I remember seeing the combat report of a pilot of a P47 (Thunderbolt) that had been ground strafing a Japanese position in Burma, where he reported that, during the attack, the (Pratt & Whitney) engine started to run roughly and upon reaching his airfield they found a cylinder completely missing (gone) and the crankcase with many bullet hits. A liquid-cooled engine with a hit in the cylinder jacket, or the radiator, would have demanded a more or less immediate bale-out by the pilot. America, during the war, largely depended upon the air-cooled engine for both bombers and fighters, with the Packard-Merlin confined to the Mustang and the Allison in the Lockheed Lightning and the early Mustangs.

One feature of American piston engines was their outward cleanliness compared with British built engines, due to a more or less complete absence of oil leaks. This was because greater attention was being paid to the quality, sealing and finish of all face joints. I recall the pristine condition of the Packard-Merlins in the Canadian built Lancasters and Mosquitoes upon delivery here after flying across the Atlantic. No oil anywhere.

This led me to ask Rolls after the war, when they were testing the civil Merlins for the Canadair North Star commercial airline aircraft, to paint their engines white for all test bed running. But this didn't work because the paint sealed the joints! So we eventually settled for a bare, scratch-brushed aluminium surface.

* * *

Sometime in the Spring of 1943, there was a meeting at the

Ministry of Supply to discuss Motor Transport (M/T) fuel, its quality and supply. The particular point was that, with the build up of U.S. and Allied forces in the U.K. in preparation for the eventual landings in Normandy, there had to be a standard M/T fuel for all the armies. To cut a long story short, it was decided that this fuel would be a leaded one of 80 octane. Two or three months after its distribution and use in the U.K. there was a spate of exhaust valve trouble in the British equipment. Another big meeting at the M.O.S. was held to decide upon the remedy. It was chaired by the technical head of M/T, Hugh Fulton, who had been technical manager of the Albion Motor Co., Glasgow, and who eventually became managing director of that company. Our friend Errol Gay of Ethyl Corporation, by then in the U.S. Army, came over from the States and I was asked to sit in. The end result was that proper advice on design and suitable valve steels was given to the truck manufacturers. The day was saved and our transport was brought into line with that of the U.S. Forces, largely due to the efforts of Fulton and Errol Gay. Jointly they arranged for Chevrolet truck valves and guides to be shipped over in quantity and for British engines to be adapted to take these, thus quickly and effectively getting production moving – all rather irritating to those of us who had been warning the British manufacturers and the Army long before the War started that T.E.L. would be added to the fuel.

*   *   *

It has already been mentioned that the production cost of the Merlin, assessed at those plants building it only, with no added development costs, was £1 per brake horsepower and £1 per lb, dry, engine weight. This compared with that of the Bristol 'Hercules', sleeve valve engine of about £2 per brake horsepower and £2 per lb of engine weight, or twice that of the Merlin. The reason for this was due to the basic difference between the two engines, liquid and air-cooling. An air-cooled engine has a larger cylinder capacity (and generally more cylinders) than its liquid-cooled counterpart, since its specific power output is lower than the latter. Furthermore the Hercules crankcase, cylinders and sleeve valves were forgings and machined all over, whereas the Merlin carcase (cylinder blocks and crankcase) were castings needing considerably less machining. We in the U.K. appear to have struck a nice balance in using these two types. Whether or not the complex Sabre engine, with its two geared crankshafts and twenty-four, sleeve valve cylinders, was worthwhile, or if the simpler Rolls-Royce Griffon would have been the better buy, is hard to say, even with post-war hindsight. For the Sabre was cleared by Type Test in 1939, well before the Griffon, and already assigned to an airframe – the Typhoon. But in the war, if the Germans had further developed their larger engines, we would

My friend of many years Lieut. General Jimmy Doolittle (Hon. K.C.B.).

have had to bring forward the Griffon earlier for the Spitfire. However, this wasn't so and the Sabre/Typhoon was given an important niche as a high performance, low level, fighter and ground attack machine. Nevertheless the Sabre, was at least four or five times more expensive to produce than the Merlin, and it was not all that efficient. The pistons and sleeve valves, when withdrawn into the crankcase during the instroke, demanded a lot of oil sloshing around inside the crankcase to lubricate and cool, which in turn demanded a large oil cooler. So, while the Sabre had a superiority in piston area (a larger number of cylinders) its mechanical efficiency was lower than either that of the Merlin or Griffon.

[1] Major G. P. Bulman, in a lecture before the RAeS some years ago, gave a good character study of Frank Halford. He said, 'Halford had no engineering training at all, none of the academic qualifications normally expected. He was the essentially creative artist, anxious to "get-on-with-the-next", and a little too apt to leave his devoted staff the drudgery and sweat of carrying his "latest-but-one-design" into production. He had immense magnetic and dynamic charm, but his mind was ever busy with his engines.'

[2] *British War Production* by M. M. Postan, H.M.S.O., 1952.

[3] *Planning in Practise, Essentials in Aircraft Planning in Wartime* by E. Devons, Cambridge University Press, 1950.

[4] The process of sintering in this case involved taking aluminium oxide powder and firing it in a kiln. China Clay was used as a fluxing agent to reduce the temperature to around 1650°C, lower than that needed to fuze pure aluminium. The clay acted as a binder.

[5] Source: *Avro aircraft since 1908*, by A. J. Jackson, Putman.

18 *The Poppet and the Sleeve Valve*

There was, particularly in the between-wars period, much argument as to the relative merits of the poppet and the sleeve valve. Fedden (Bristol) eventually produced a successful sleeve valve after some twelve years of intensive development but it could be argued that the Jupiter, Mercury and Pegasus poppet valves were relatively crude examples of the art and were poorly cooled and only nominally lubricated. A true comparison would be between these poppet valve examples and the two American engines, the Pratt & Whitney and the Wright Cyclone, which had superior cylinder head finning and N.A.C.A. cowling (for cooling) together with continuous oil circulation through enclosed rocker boxes. The Americans used two valves instead of four and that helped to cool the valves by increasing the fin area around the single exhaust valve. To sum up: the sleeve valve was more expensive than the poppet valve to produce and did not give better engine performance. It was always claimed by Fedden that the sleeve valve cylinder could use a higher compression ratio than its poppet counterpart. But this was misleading, since the sleeve valve gave inferior volumetric efficiency (cylinder filling) than the poppet and could, therefore, accept a higher compression ratio. It also lost something in mechanical efficiency. As power (boost) was increased there was, with Bristol sleeve valve engines, almost continuous development of the junk head (the head that slipped in-between the sleeve and the cylinder proper) to control oil leakage past the junk head rings and also prevent them sticking, though the final type of junk head, in copper with a steel insert for the rings, was a great improvement. There was only one direct or comparative test between sleeve and poppet valve engines that I was able to initiate. This was done on a single cylinder Merlin unit and that of the Eagle engine. The Eagle (reviving a famous name) was similar to but had a larger cylinder capacity than the Sabre, also having twenty-four, sleeve-valve, cylinders and geared double crankshafts. The tests showed that while the Eagle cylinder had a higher compression ratio than the Merlin, it gave no more power or better fuel consumption. This was the only strictly comparative test done between the sleeve and the poppet valve. I would say, therefore, that with proper valve technology, i.e. a sodium cooled, hollow head, valve and circulatory oil lubrication and superior cooling of the valve stem, air-cooled engines such as the Pratt & Whitney and Wright, were more economical to manufacture than any of the equivalent Bristol sleeve valve types.

Of the Merlin installations in those aircraft designed to take the engine (Hurricane, Spitfire, Lancaster, Mosquito), the most successful engine exchange was that of the Merlin in place of the Allison in the Mustang. This fighter (first delivered here in April 1942) was designed by the North American Aviation Corporation of Los Angeles to British operational requirements and fitted with the liquid-cooled, V12, Allison. We had ordered some 600, though I do not believe that the full number was delivered. The Allison, being a low altitude engine (a low full throttle height supercharger) made the Mustang (as Harker records in his book, *Rolls-Royce From the Wings*) suitable only for Army co-operation and photo-reconnaissance duties. But this was a considerable limitation of a potentially magnificent aeroplane and, after Harker (a Rolls-Royce test pilot) had flown this first example (AG-422), he immediately saw its possibilities as an air-superiority and long range penetration fighter if fitted with the 'Merlin 60' series engine which had a two-speed and two-stage supercharger. After this flight, through the courtesy and far-sightedness of Ian Campbell-Orde, the Commander of the Air Fighter Development Unit (AFDU) at Duxford, Ronnie Harker had difficulty in persuading his own colleagues and some senior staff at Rolls that such a change would be worthwhile and warranted. But, eventually, with help and advice from (the late) Ray Dorey who was in charge of the Rolls engine flight test section at Hucknall, Harker eventually put his case to Hives, personally, to whom he had previously written a note. After the talk, Hives acted quickly and saw my Chief, Sir Wilfrid Freeman, who immediately arranged for three Mustangs to be sent to Hucknall for the Merlin conversion. Not only did the Mustang become the air superiority fighter of the war but, with long range fuel tanks, it could even escort the U.S. Eighth B17s to Berlin.

19 From Pistons to Jets

It already seemed in 1944 that the pure turbojet would be the engine for fighters, but it was then uncertain whether or not it could eventually be made sufficiently economical for long range bombers and reconnaissance aircraft. For instance, on the commercial side, the Proteus propeller turbine of about 4,000 eshp had been chosen for the Bristol Britannia. It was a two shaft engine with a free turbine driving the propeller. At about the same time, the Air Ministry were considering the post-war replacement of the North American Harvard trainer and had decided upon a propeller turbine instead of a piston engine. I pointed out that, whereas a trainer airframe could be designed and produced in prototype form in about eighteen months or two years, a propeller turbine would probably need four or five years before it could be ready and, therefore, the timing between the two would be astray. However, it was decided to go ahead and two aircraft were ordered, the Athena from Avro and the Balliol from Boulton & Paul. We also ordered two engines, the Rolls-Royce Dart with a two-stage centrifugal compressor and the Armstrong-Siddeley Mamba having an axial compressor. It was thought the former would stand more rough treatment than the relatively delicate small axial, particularly from grit and foreign object damage.

All this showed that the various engines which were evolving during the mid-1940s from Whittle's original concept, were already ousting the large piston engine in aviation.

During my service as Director of Engine R. & D., our objective was to plan the post-war future of the aero gas turbine for both military use and commercial applications. For students interested in the history of the technical side of engine development, no better book has been written than that referred to below.[1]

I laid down the policy then that, in order to establish a 'state-of-the-art', we would order two engines for each new aircraft requirement. It was very desirable for the engine design groups to compete with each other and build up a good background of experience. This would not be practicable today, with present engines of four, five and even eight times the thrust of those at the end of the war, and with cost escalation unfortunately more than keeping pace.

In the case of the Dart and the Mamba our enthusiasm was not so much for the trainers, since a single propeller turbine did not, in our minds, represent what a Service pilot would eventually handle. He would fly either a single or twin turbojet fighter or, perhaps, a bomber or reconnaissance aircraft with four turbojets or propeller turbines. We were, however, eager to undertake the development of these two engines because we thought there was a use for them in commercial aviation, for a feederliner or medium range aircraft. Funds would not have been readily forthcoming in 1945 for other than a military requirement. In the event, the Boulton & Paul Balliol became the chosen trainer for the Royal Air Force but was fitted with the Merlin.

Working on the Brabazon Committee recommendations for post-war commercial aviation, Vickers (now B.A.C.) designed a four engine medium range airliner, the Viscount, fitted with the Dart. Armstrong-Whitworth built the Apollo prototype for the same purpose, powered by four Mambas. The Viscount won out and became the biggest commercial production success of the post-war British aircraft industry, some 450 being built and sold worldwide.

When I went to see Hives of Rolls-Royce to tell him that he would get a contract for the Dart he didn't seem very enthusiastic. I wanted him to have it designed and developed at Barnoldswick where knowledge of turbine design was becoming well established under Hooker and Lombard. But 'Hs.' told me that he wished to concentrate all post war engineering at Derby and wanted the Dart there. I wasn't too happy, but he was in charge and one could hardly argue except to point out the engine's importance to us at the Ministry. Hives thought that the Dart was an engine that Derby could cut their teeth on, though I suspected at the time he hadn't too much enthusiasm for the engine. He was, of course, right in the long term; all engineering had eventually to come back to Derby. But, in the event, the prototype Dart came out overweight and under power. I had left before the engine was ready to run, and Maurice Luby, my successor who had been my deputy, came to see me to ask if I were in his place would I scrap the Dart in favour of the Mamba (which was doing quite well in its early running). I said, 'No, of course not; Rolls will get the engine right and it will probably go into the correct (most saleable) aircraft'. And so it did. The Apollo went no further than the prototype, and the Mamba was fitted in the Fairey Gannet for the Fleet Air Arm. This was a twin-engine arrangement, with two engines connected to a gearbox each driving separately one half of a co-axial, contra-rotating, propeller.

Another very important engine requirement coming at about the same time was that for the Mosquito replacement. This was strictly a Ministry initiated turbojet requirement placed by us with Rolls-Royce for the AJ65 (6,500 lb thrust), later to be named the Avon, for the English Electric Canberra bomber designed by (the late) Teddy Petter. The AJ65 was also the first Rolls axial compressor type engine to go into production and, following our practice of duplicating every new engine requirement, we looked around for a likely counterpart and competitor. Before any decision had been made George Watt came into my office one day and said that he had Drs Baumann and Smith of Metrovick in his office, trying to sell him an uprated Beryl. I said that I didn't want an uprated Beryl, but if George would bring them in we would have a talk and decide what they should build.

Metrovick's gas turbine department was very good on axial compressor design and I put it to Baumann and Smith that we wanted a competitive and complementary engine to the AJ65, but it should be of 7,000 lb/thrust and have a better specific fuel consumption (sfc) than the AJ65. Baumann asked why we were making it more difficult for them than for Rolls, and I told him that Rolls were the more experienced in aero engine development whereas Metrovick, though more knowledgeable in turbine technology, might be slower than Rolls to get their engine into service and by that time more thrust would be wanted. It was finally agreed that they would get a contract for, I think, six development engines of a basic design to give 7,000 lb/t and about 0.78 lb/lb/t sfc. The engine, originally known as the F9, was later named the Sapphire. Since there was the need at the war's end to reduce the total number of firms engaged on aero engine work and because those steam turbine manufacturers brought in early on the gas turbine might wish to concentrate on industrial gas turbine engines the Sapphire was handed over to Armstrong-Siddeley to develop. They did a considerable redesign of the engine mechanically but, of course, retained Metrovick's compressor and turbine with their good aero- and gas-dynamics. They fitted Syd Allen's (of Armstrong Siddeley) 'walking stick', annular, vapourizing combustion chamber for which we had already backed Armstrong with a development contract. The aim was to reduce engine diameter and to improve the temperature pattern at the first stage nozzles ahead of the turbine.

This change of firms came after I had left. Frank Halford had made a bid for the Sapphire for de Havilland but, without warning, the Ministry switched it to Armstrong-Siddeley. Frank felt very strongly about this at the time, but nothing could be done. The Armstrong team was never very good at intensive engine development and this went back to John Siddeley (the first Lord Kenilworth) who would only agree the minimum to get engines out of the factory. Had this been otherwise the Sapphire might well have taken the place of the AJ65 in some of the more successful aircraft. Later, Rolls-Royce were to negotiate the Sapphire's compressor design for the RA14, their new mark of Avon.

During this period (1944–1946) the following engines were either in production or in development: from Rolls, the Derwent, Nene and Tay (all double-sided centrifugal compressor types, of 3,000 lb/t, plus, to 5,000 lb/t and 6,000 lb/t respectively) with the axial compressor Avon (AJ65) of 6,500 lb/t; from de Havilland the Goblin and the Ghost (of 3,000 lb/t and 5,000 lb/t) both of which had single-sided centrifugal compressors and were, therefore, of larger diameter than the Rolls-Royce engines. In fact, when I took over Engine R. & D., the Ghost had already been ordered, but its diameter (57″) was too large so I asked Halford to reduce it. That is how the comparatively large number of separate combustion chambers was reduced to fewer of somewhat greater diameter, paired to a common breeches elbow from the compressor diffuser casing, which reduced the overall diameter of the Ghost II to about 54 in. Bristol produced two propeller turbines, the high economy Theseus of 2,500 eshp, with heat exchanger, and the Proteus of over 4,000 eshp.

Before I left in 1946, we had already discussed with Bristol Frank Owner's two-shaft, axial turbojet design. It was later ordered as the Olympus and redesigned by Hooker and the Bristol team for the Avro Vulcan bomber. Also from Rolls-Royce was to come the low bypass ratio ducted fan, suggested by Griffiths and subsequently developed by Lombard as the Conway. Even in that short time of about two years, the production turbojet had grown from about 3,000 lb/t to 7,000 lb/t. Today, nearly thirty years later, we see turbofans of high bypass ratio (5 and 8:1) producing 45,000, 50,000 and even 60,000 pounds of thrust and specific fuel consumptions of less than one half of those early engines.

There was the formation of a very important Committee in M.A.P., initiated by Roxbee Cox (now Lord Kings Norton) on the 3rd October, 1941. This was the Gas Turbine Collaboration Committee which, although purely domestic, was known to our friends in the U.S.[2] It covered two periods: one ending and one beginning in the autumn of 1941. The earlier period was characterised by associations of pairs of teams with clear-cut common and limited objectives such as Power Jets and British-Thomson-Houston's collaboration on the early Whittle engines and the Royal Aircraft Establishment and Metropolitan-Vickers on axial compressors with Power Jets and the Gloster Aircraft Co collaborating. But, to ensure economy of effort, Roxbee Cox, in the second stage, proposed the formation of the Gas Turbine Collaboration Committee, to include *all* firms engaged on aero gas turbine work with the official bodies such as the Admiralty, the Royal Aircraft Establishment (R.A.E.) and National Physical Laboratory (N.P.L.). This Committee did exceedingly valuable work throughout the war period, where the firms discussed their problems and the measures taken to solve them. It faded out sometime after the war since competition naturally entered into things.

When the European War was about at its end and the Allied armies were already into Germany, we and our allies had working parties consisting of engineering, technical and scientific personnel, all temporarily in uniform, seeking out German wartime developments in the aeronautical, engine, equipment and armaments fields. We asked (the late) Sir Roy Fedden to head one group and he did a magnificent job in unearthing gas turbine and rocket propulsion work, helped by some of our staff. I also had one or two personnel in France. There were other groups, headed by senior people from the R.A.E., Farnborough and the firms, investigating aerodynamic developments and rocketry.

I was 'phoned one day by a member of my staff in Paris, who asked if I knew someone by the name of Coanda. I did. It was Henri Coanda, a Rumanian who had lived many years in France. He had in fact been engaged as an aircraft designer by Bristol before the First World War. Coanda had a small man and boy size aerodynamic laboratory in Paris in which he was continuing to investigate his 'Coanda Effect' for various applications. The Germans had visited him during the Occupation. However, immediately after the liberation of Paris, there was a good deal of paying back old scores between the local French Resistance and other citizens. The former, Communist dominated, accused others of being collaborationists and hauled them off to prison; even, it was said, shooting them out of hand. Coanda was incarcerated because someone had seen Germans visiting his laboratory. Since he could well have been in imminent danger of being shot, I told my colleague in Paris to get in touch with the military authorities and have Coanda released for interrogation. Luckily, this was done but not before someone from Farnborough (I forget who), who knew of his work, had been to Paris to confirm my request.

I went to Germany to see the Hermann Goering Research Institute at Volkenröde, near Brunswick. Our Army had arrived there and that evening, when I was dining with a Colonel, he told me that they had been to the Belsen concentration camp that day and he was appalled at what they saw. He said that everyone should see it to appreciate how low human behaviour had sunk, and offered to drive me in the next morning. I didn't quite know what to expect, and was nearly physically sick when I saw it all. The Army were starting to clean it up, with bulldozers heaping up wax-like corpses, almost skeletons, and mechanical diggers making mass burial pits. Some of the living could just about walk while others were lying about awaiting attention from our medical units, their eyes seemingly staring at nothing and with completely vacant, expressionless, faces. The sight of these and the awful stench was something that one couldn't describe, only experience. I heard that, in other concentration camps overrun by American troops, some of the German guards who hadn't deserted and run away were summarily shot by the Americans. I thought I had experienced everything after seeing the results of Bolshevik atrocities in South Russia in 1919, but it seemed almost inconceivable that Belsen represented deliberate Government policy, as distinct from the Russian case of unruly hordes of low grade humanity taking it out of those on the other side.

In 1945, just after V.E. Day but before V.J. Day, I was invited to the United States of America by the U.S.A.F. to discuss with them and the aero engine manufacturers our views on engine development and the future of the aero gas turbine. I left from Poole Harbour in a Boeing 314 flying boat piloted by a senior B.O.A.C. Captain and a friend of mine, Gordon Store. We flew first to Foynes in Southern Ireland, to top up with fuel for the Atlantic crossing, and landed at Baltimore early the next day. It was a comfortable flight, particularly so because, unlike a landplane of that period, one could walk about in a flying boat. Then from Baltimore I went by car to Washington and the offices of the British Mission. The plan mapped out for my tour was very extensive and intensive, including visits to the U.S.A.F. in the Pentagon and the Navy Bureau of Aeronautics Power Plant section. There I met my old friend Carl Fliedner, the civilian technical adviser to the section and 'father confessor' to many generations of Naval Bureau chiefs. I was loaned a Beechcraft C45 by the U.S.A.F., and had an R.A.F. Squadron Leader with a Flight Lieutenant seconded for piloting duties, with an American top sergeant crew chief to look after the day-to-day servicing of the machine.

We flew across and round the States, calling at all the principal aircraft constructors and aero engine manufacturers, including those industrial firms that had been brought in on gas turbine work, such as Allis Chalmers, Westinghouse and General Electric. Early in the war G.E. had been given the Whittle designs by General 'Hap' Arnold, C.A.S. of the U.S.A.F. who, in turn, had received them from Lord Beaverbrook.

At Muroc dry lake, now Edwards Air Force Base, the experimental flight test centre on this side of the Sierras from the West Coast, I flew my first turbojet aircraft, the Bell Aircomet, P159-A, fitted with two General Electric I-16 engines each of about 1,600 lb/t. It was the first jet-propelled aircraft built in the U.S., and had a top speed of about 400 mph. Following this, I flew the Lockheed Shooting Star P80A, with a General Electric 1-40 turbojet of around 4,000 lb/t. It had a maximum speed of about 550 mph and was nice to handle. Its prototype, the XP80, had been designed, built and flown by Lockheed in the space of 143 days and was fitted with the Halford (DH) H1 turbojet.

While visiting Boeing at Seattle we were shown a bomber mock-up they were preparing for, presumably, a U.S.A.F. requirement. It was a high wing design plastered with engines about its centre section in closely spaced groups of four or six G.E. turbojets, then of relatively low thrust. While we were there George Shairer, the Boeing design head, who was then with a working party in Germany, cabled home to tell Boeing to hold what they were doing since he had come across something better. What he had found was the swept wing and the podded engine, and this combination resulted in the Boeing B47, America's first long range turbojet bomber, which first flew in December 1947 and led to the bigger and better B52.

We were flying across the States, bound for Los Angeles from Washington D.C. The first atomic bombs had been dropped on Japan and all the broadcasting stations were buzzing with the news of the impending Japanese surrender. We landed at Mines Field (now Los Angeles International Airport) exactly at 4 p.m. West Coast time on the 14th August to be greeted by the sirens and car hooters blaring at the news of V.J. Day. We were directed to taxy and park alongside a B29 bomber. As with all our stops for refuelling at military bases across the States, the punctilious and courteous attitude of the U.S.A.F. personnel was most marked and pleasant. Our pilot would be asked by the control tower to report who was on board from a Colonel to one-star General upwards. Then the base commander, usually a full Colonel, would meet us. So it was at Mines Field, with the commandant coming to greet us. When I had shaken his hand and thanked him, I made a little joke, quite spontaneous as it happened; pointing to the B29 I said, 'That's a bit out of date now'. This really roused the poor man, who proceeded to tell me what a wonderful machine it was and what it had done.

That V.J. Day evening was 'dry', all bars in Los Angeles being closed, so I had to look up old friends known to me before the war and hope that one or two who had not left during hostilities might come to our rescue. We were not disappointed. There was that night an interesting insight into the American attitude to war and peace;

by midnight, every aircraft manufacturer had received a cable from Washington stopping all production contracts. The many thousands of people who had left work that afternoon only came back to collect their pay.

In L.A. I again met 'Dutch' Kindelberger and Lee Atwood of North American, also Ray Rice and Ed Schmued who were largely responsible for the successful design of the P51 Mustang; then Robert (Bob) Gross, boss of Lockheed, and the inimitable 'Kelly' Johnson. It was all a magnificent reunion. At the Douglas El Segundo plant, I met for the first time that charming and very capable aeronautical engineer, Ed Heinemann, who was responsible for the design of some of the most successful U.S. Navy fighters.

During our stay of some ten days in Los Angeles I was at my favourite Beverly Hills Hotel and one morning joined Albert Plesman (head of K.L.M.) for breakfast. We had a long talk, from which I gathered that he was seeing Douglas with plans for building up, or rebuilding, his commercial fleet. He mentioned how his energetic staff were already sorting things out at Schipol under his capable lieutenant, Veenendaal (tragically killed later with Parmentier in a Constellation at Prestwick). Later that week I met Donald Douglas Snr. for lunch with Plesman, and we discussed the future possible application of the turbojet in commercial aircraft and the likely time scale.

One day a friend who was also a personal friend of Basil Rathbone said that the great film star was giving a big party and had heard that there was a Royal Air Force officer in town and would he invite me. The party, held at Rathbone's house on the Bel Air estate, was originally for 60 guests but Basil's wife, his second and a most charming woman, kept asking any friends she met casually and it finished up with at least 90 people. All those that mattered in the film and artistic world were there: Myrna Loy, William Powell, Artur Rubenstein, the famous pianist, who later played for us, Ethel Barrymore, Mary Pickford, Gary Cooper, Cary Grant and so on and so on. Cocktails went on from 6.30 p.m. until 9 p.m., so everyone was fairly 'high' when we sat down to dinner. I was placed between Mary Pickford and Ethel Barrymore. The former kept on telling me about someone called 'Buddy' and his war work in the Pacific. Only later did I realize she was talking about her husband, 'Buddy' Rogers, the young man she had married after parting from Douglas Fairbanks, Snr. Half way through the evening it poured with rain and we all had to come back into the house and the party became rather wild. There was more drinking before break-up at 4 a.m., and back to bed. It was the best dinner, so far as the food was concerned, that I had enjoyed for a long time. Con, my wife, when very young, lived in Newport, Wales, and used to cycle with Basil Rathbone. I mentioned this to him and he was very pleased to be reminded of his time there. Rathbone had his dramatic training at Sir Frank Benson's School.

I then visited the Wright-Patterson Base at Dayton, Ohio. Eddie Page had died and Opie Chenoweth was in charge at the time. A general discussion ensued, mostly about the gas turbine's future, with the swapping of views on the difference between British and U.S. procurement practices and engine development. Our procurement of engines was, I think, superior at that time to the American. It got under way more quickly and more engines were ordered at the start of a contract to get things moving. The mutual trust between the British Ministry and the engine manufacturers was good and involved less paperwork than the U.S. system. Today, I would say there is little difference between our basic methods of procurement and those of the States, though my impression is that the Pentagon still demands a massive amount of paperwork compared with our own Government Departments.

Today the quality of design and the development of the aero gas turbine in both countries is very good, but the U.S. have the advantage of an immeasurably greater domestic market for both military and commercial aircraft than the U.K. Therefore they can acquire a better statistical appraisal from large numbers of engines in service, which makes for more economical maintenance and rapid modification procedures.

There were, in the U.S., as in the U.K., some industrial steam turbine manufacturers (for public utility electrical generation and marine propulsion) who, it was thought, might usefully help in the evolution and development of the aero gas turbine. But this did not necessarily prove to be so. Admittedly, the legitimate piston aero engine builders were at that time too fully involved in the war to take on the gas turbine energetically.

The real success of the aero gas turbine and its rapid progress from Whittle's original concept was due to the background knowledge built up with the piston engine and its development in the between-war years, and during the war. This resulted in high quality engineering, improved materials, manufacture, inspection and reproducibility. It would have taken the industrial engineering firms many years to do what the aero engine manufacturers had achieved within two or three years of seeing the Whittle engine. This is all understandable when it is realized that, except perhaps for marine propulsion, there is little series production in the case of turbo-alternators for electric generating stations. Few are even alike in detail and all are built to greater safety factors than aero engines. Weight matters little, whereas both airframe and aero engine, having to fight gravity, must be as light as practicable.

Recent years have, however, shown how effective an already developed aero gas turbine is for naval ship propulsion and for 'peak lopping' duties in electrical generating stations. Many are now put to such use and, more recently, for high capacity gas pumping purposes.

General Electric of America were the exception in the industrial field since, during the war, they had manufactured many thousands of exhaust driven turbo-blowers for aero piston engines at their Lynn plant. The turbocharger was developed before the War under the aegis of the fiery Dr Sanford Moss and the U.S.A.A.F. at Wright Field. G.E. had, therefore, acquired sufficient knowledge and experience needed to enter the aero gas turbine field. When after V.J. Day I visited G.E.'s headquarters at Schenectady in New York State I met the boss, 'Electric Charlie' Wilson who, at lunch with others, asked me if I thought General Electric should continue in the aero gas turbine business post-war. Since they had already built the I-16 and I-40 gas turbines derived from the basic Whittle designs I said that I thought they should go ahead. I also added the rider that, to be successful, they must divorce the aero gas turbine business completely from that of their steam turbines, as it was a different discipline and, anyway, the salary scales would be higher than those paid in the industrial field. I don't suppose for one moment they acted because I said this, but it merely confirmed what they had already made up their minds to do. Their plants at Lynn and Evendale now produce engines of the highest quality and performance in aero gas turbine technology, brought to this level under the control of people such as Jim LaPierre, Jack Parker and Gerhard Neumann, with a whole group of capable technicians and top manufacturing specialists. Remarkably, General Electric aero gas turbine facility only employs about 25,000 people, compared with Pratt & Whitney's 45,000 and Rolls-Royce's 63,000. But Rolls do much more 'in house' manufacturing than the two American firms. Pratt & Whitney have achieved their position in the aero engine business over many years, from which they have a steady production flow, spares alone representing some 50 per cent of their total business.

In contrast, at that time (1945), Westinghouse, whose first shot at a turbojet, the Yankee or the 19B (19 in diameter), was very successful, got lulled or gulled into thinking they had the answer to their aero engine future. They hadn't. They were charming people but didn't seem to realize that it was not feasible just to put one of their senior industrial department managers in charge and get results – even if there was a capable man heading design. But there was little chance of recruiting further high level aero engine technicians when the salaries had to conform across the board with those on the industrial side. So Westinghouse eventually faded out of the aviation field though they have since been very successful in other highly technical developments, such as the application of nuclear energy to public utility electric generation.

Visiting the 'old troupers', Pratt & Whitney, and talking with my good friend L. S. (Luke) Hobbs, was very refreshing. Luke wasn't born yesterday and he spent the two days' visit in plying me with questions about our views on the gas turbine's future. He, with his project and design chief, Perry Pratt, already had ideas about what they would do and indeed eventually came to the two-spool axial engine. At first I think they had decided upon it as a large propeller turbine but, ultimately, and at about the same time as Owner of Bristol thought of his two-spool Olympus, Pratt & Whitney produced the very successful J57.

There was, later, what might be called the 'Nene and Tay' episode. This developed when Phil Taylor, Vice-President Engineering, of Curtiss Wright, and S. T. Robinson, both left Wright and, presumably in anticipation of the turbojet's future, did a deal with Hives of Rolls-Royce for the manufacturing rights of the Nene and the Tay in the U.S. To shorten the story: the U.S. Navy was interested but wished Pratt & Whitney to take over the engines, which was done, resulting in a very good deal for Taylor and Robinson. It was alleged afterwards that P. & W. thought Rolls had sold them the Nene and Tay in a half-baked or underdeveloped state; but considering the Nene at that time had only a limited background in the Fleet Air Arm Hawker 'Sea Hawk', and the Tay had had no British application, both were bound to need considerable further development. This was effectively done by P. & W.

There was an amusing incident when we were flying back East, across the States. We had put down at Billings, Montana, to refuel for the next stage on our way to Milwaukee. Billings was the birthplace of the famous Western cowboy film star, the late William S. Hart, and he is remembered at the airport by a bronze statue of him and his favourite horse. We went into the small canteen for bacon, eggs and coffee, which was ordered by the second pilot who had a very Oxford accent. Upon hearing it, the little girl behind the counter taking the order looked up at him and suddenly burst out laughing and rushed into the kitchen. Then, of course, we had to tell her, and the now assembled company, who we were and later retail to them what the European War had been like and how life was in England.

Arriving in the late afternoon at Sioux Falls in South Dakota, near the border of Nebraska and the last stage on our flight to Milwaukee, we went to the hotel where we had been told there were reservations, only to be informed by the room clerk that they hadn't heard about us. Sioux Falls was a large assembly place for the Army, and it was literally overflowing with troops who were about to be demobilized or sent elsewhere. The room clerk said, however, that we could have a small room for all four of us. Then the Squadron Leader drew the clerk aside and told him I was a 'General in the British Royal Air Force' (not emphasizing my single star). The clerk suddenly reacted and gave us the Presidential suite, where there was one large bedroom and another bedroom with twin beds for the pilots. The small room originally offered was given to our top sergeant.

Our next stop, Milwaukee, is famous for brewing and beer, an activity started by Germans who emigrated there. Along with Cincinnati and St Louis it was, before World War I, known as one of the Kaiser's three American colonies. It has also been called 'The Cream City' because it is largely built of very attractive light yellow bricks. A famous citizen was Air Corps General Billy Mitchell, a protagonist of air power, which he demonstrated practically in Chesapeake Bay, in mid-1921, by sinking some old U.S. Navy ships; a submarine and a destroyer, and the battleships Virginia and New Jersey and afterwards the ex-German cruiser Frankfurt and the heavily armoured and constructionally sub-divided battleship the Ostfriesland. For the former, 500-1,100 lb bombs were used the ex-German being sunk with 2,000 lb bombs. Mitchell got himself into trouble with the Navy and also the Washington Government for being, like many other evangelists before and since, too vocal or political in trying to gain his ends. Naturally, the U.S. Navy didn't wish to have its budget cut or be 'sunk' without a fight.

My Milwaukee stop was to visit the well-known industrial firm of Allis-Chalmers and exchange views on gas turbine developments. They proved very friendly and most capable people but it seemed that they wished to return to their industrial work, which indeed they hadn't really left.

Allis-Chalmers had been studying the design of a turbo-driven ducted fan in 1943 for the U.S. Navy, but nothing came of this and the Navy cancelled the contract. Both the U.S. Army and Navy were keen on getting the de H Goblin (3,000 lb/t) produced by Allis-Chalmers, but this project again fell through when an engine of greater thrust was already becoming available, i.e. the GE I-40 of 3,750–4,000 lb/t.

On our way back to Washington I went to Detroit to visit 'Boss' Kettering, Vice-President of Research of General Motors, and to see my Detroit friends at Ethyl Corporation. Then on to New Jersey to visit Wright Aeronautical (Curtiss Wright) at their wartime built, windowless, plant in Wood Ridge. There was an uneasy atmosphere at Wright since Guy Vaughan, the President, was trying to grasp the firm by the neck and, in so doing, only making things worse. In this immediate post-war period under Guy they suffered badly and good engineering heads like Arthur Nutt, Phil Taylor and Ray Young who followed him, left. Guy came to England some time after I had returned and asked me if I could advise him on a manufacturing licence for a gas turbine. He mentioned the D.H. Ghost turbojet (5,000 lb/t) and, possibly, the Venom fighter for building in the States. But I advised against it, if only because there were more advanced engines already in development and nearly ready to run – the Avon and Sapphire. Eventually Guy retired and Roy Hurley (of Hurley-Townsend sparking plugs before the War), then with Breech at postwar Ford, became President. He took the licence for the Armstrong-Siddeley Sapphire and the Bristol designed and modified Olympus, called the Zephyr, which was intended for commercial airline aircraft. The Wright built Sapphire (J65) was a success in the Martin B57 (Canberra) and also in the Republic F84F fighter. The Zephyr didn't go beyond the test bed because Hurley tried to sell it to the West Coast aircraft builders before it was fully developed.

At the end of my round of U.S. visits I arranged to fly to Trinidad to see the hydrogenation plant for 100 octane fuel which had been built at U.K. Government expense and operated by Trinidad Leaseholds at their Pointe-à-Pierre refinery. My late brother Jack (B. G. Banks) managed the refinery and had been in Trinidad for some years, so I was glad to have the opportunity of seeing him and his wife Helen and family.

The flight from Florida consisted of two long hops, the C45 already having an extra fuel tank in the nose section. We flew via Nassau in the Bahamas and Puerto Rico. This first stage, and that from Puerto Rico to Port of Spain, Trinidad, literally stretched the aircraft's range to the maximum, particularly since we had on a number of occasions to circumnavigate some really heavy rain and thunderstorms.

A somewhat irritating, but in retrospect amusing, moment came when we landed at Borinquen Air Force Base at Aquadilla in Puerto Rico. Since the War was over, V.J. Day and all, the authorities there had instructions only to refuel non-U.S. aircraft for cash! This hadn't operated in the U.S. since, actually, the aircraft we were flying *was* U.S. property; but the Borinquen people seemed to think that while it might be a U.S. aircraft it had a British crew and passenger, therefore it must be 'lease-lend' equipment. Incidentally, it was the only U.S.A.F. Base in all our round trip where the C.O. didn't come and formally greet us but this was possibly due to our landing in a formidable storm after the control tower had told us to delay the approach until it passed over. Unfortunately, we were at the (fuel) limit of our range and had to land. The torrential rain was too heavy for us to disembark, but like all those late summer storms it passed over fairly quickly. Eventually, the flight crew, all innocent and unprepared, went to make arrangements for refuelling and accommodation for the night, as it was too late to start for Trinidad. They came back to the machine rather downcast and gave me the news about paying for the refuelling. I was just about to start to reconnoitre and find someone in authority when a passing jeep stopped, a U.S.A.F. Captain stepped out and asked who we were and where were we bound for. He then suddenly realized that I was a 'one star general', which was indicated on the machine anyway. Then things began to happen. We were immediately taken to the Base Commander, a very charming Colonel, who was most embarrassed since, as I have said, the Americans do take matters of protocol seriously. Signals went to Washington and within a few hours everything was cleared. I was given a General's quarters and we were well entertained to dinner and talked into the night on the European and Japanese Wars. We left early next morning for Trinidad and arrived at Port of Spain about midday to be greeted by Jack and Helen and our namesake, Maurice Banks, who was then the Director of Civil Aviation in Trinidad.

I spent a very pleasant three or four days at Point-à-Pierre, while the crew were enjoying the 'delights' of Port of Spain – not really wildly exciting but full of history, mostly concerned with the Spanish and our pirating and early trading days. Being relatively close to the mouth of the Orinoco river, the water around Point-à-Pierre was quite discoloured and sandy or muddy. We toured the south of the island and in the forests there were the howling monkeys who set up a piercing howling and chattering. There are quite large pythons there but we didn't see any.

Finally, we took our leave and flew back to Washington where I bade farewell to my faithful crew and very reliable aircraft, which had flown us some 8/10,000 miles without the P. & W. engines once missing a beat.

I gave a cocktail party at the Carlton Hotel to some of my American and British friends who had given me such generous help and made the whole trip so successful from my point of view, even though it had been at their invitation in the first place. Generals Don Putt, Pop Powers and my friend Jimmy Doolittle came along, with two Admirals, one of whom brought Sir Hubert Wilkins, the Arctic explorer, who happened to be in Washington at the time. He told me of an interesting experience before the War, when he had been invited to do a lecture tour in Japan. He was in Singapore awaiting the Japanese ship to take him to Yokohama. It was a day late in arriving, so the Company's agent, a Japanese, invited him to a Japanese club across from Singapore Island on the mainland. Most of the night was spent with a number of Japanese traders and business people of Singapore, nearly all of whom were military and naval officers in mufti. An apparently senior member told Wilkins that if and when another major European war came the Japanese would probably fight against us. Asked why, he said that when they were on our side in World War I they got nothing in return, particularly no more land to allow them to spread out and expand from the overcrowded Japanese islands – hence their invasion of Manchuria between wars. He added that even if they were on the losing side they could wait for fifty years and have another try. A truly oriental outlook.

When it was time to leave for home I could not get a flight, since the limited air service across the Atlantic was solidly reserved for V.I.P.s, so I was booked to sail from Halifax in the *Queen Elizabeth*. I went to Montreal to see my old friend and naval colleague, Paul Earl, who, like myself, had achieved the rank of Commodore, but in the Canadian Navy. There followed a long and dreary journey in a troop train, which had to pick up groups of personnel, mostly Air Force, on the way home from Canadian training bases. Arriving at Halifax, we immediately boarded the Q.E. and, luckily, I was

Reunion held at the Cloud Club, Chrysler building, New York in August, 1945. The Author with (centre) Mr Earl Webb (President of the Ethyl Corporation) and Jack Macauley. By permission of the Ethyl Corporation.

given a cabin to share with another Air Commodore. The ship was crowded to the gunnels with returning officers and men from Japanese prisoner-of-war camps, mainly those who were not in too bad health and could look after themselves; but they looked bad enough to us relatively healthy ones. Also on board was General 'Pug' Ismay and his wife. Ismay, it will be remembered, was Churchill's military adviser in the War Cabinet throughout the War.

Back home at the Ministry of Supply, the new name given to the Ministry of Aircraft Production, I found things much the same. Ralph Sorley had gone and Air Marshal Sir Alec Coryton was C.R.D. My deputy, a naval captain, Maurice Luby, had kept things going in my month and a half's absence. But they had done one thing which I wasn't too pleased about. They had given Napier a contract for a propeller turbine of about 1,500 eshp, the Naiad. We had already ordered (I had personally ordered) the Dart and Mamba; and though 1,000 eshp engines, they were capable of being stretched beyond 1,500 eshp.

I had also ordered a high economy engine from Napier, the Nomad, intended for maritime reconnaissance or long endurance operations. It was a flat, horizontally opposed, 12 cylinder, two-stroke, compression-ignition engine with sleeve valves, having a mechanically driven blower supercharged by an exhaust driven axial compressor. The Nomad was from the start a highly experimental project and we did not know at that time whether or not a propeller turbine would be able to give the minimum specific fuel consumption needed at the relatively low altitudes of the maritime rôle.

When Sammons, who was then managing director of Napier, showed me the Nomad's design, he asked me what I thought the development time to type test would be and I replied, 'Six or seven years'. He thought they could do it in 3–4 years and I disagreed, but Sammy said that he had had experience of similar engines when he was earlier with Petter Oil Engines working on Kadanacy's principle of boosting loop scavenge two-strokes by ballistic exhaust pipe tuning. The Nomad was, however, a very high duty engine and quite unlike any other in his experience.

English Electric took on Sammons as Technical Director of Napier shortly after they assumed control, when Sir Geoffrey de Havilland had asked for the complete release of Frank Halford to look after their Engine Company. Halford's gas turbines, the Goblin and Ghost, were then coming along rapidly. He, Sammons, was a good general engineer but an incurable optimist and that could lead to difficulties in respect of estimates of time and cost on a contract. This was particularly true in the case of the Nomad, admittedly a complex engine. When I came back to the M.O.S. on loan from Associated Ethyl, in 1952, it was in its seventh year with no sign of getting its brochure power or specific fuel consumption, and no nearer Type Test. In fact, almost as soon as I arrived, I was approached by Ned Dunnett, Deputy Secretary (now Sir James Dunnett), who asked me if I couldn't stop the Nomad. I said that I intended to ask Napier first to show me that it could attain its power (3,000 bhp) and, more important, give its brochure specific fuel consumption of 0.32 lb/bhp/hr. If they could not give me this practical demonstration I would certainly cancel the contract. In fact, towards the end of my loan period, they were unable to satisfy me. Before I came back to the Ministry they had already done a major re-design, and eliminated the mechanically driven centrifugal blower, since the axial did better than expected. This also considerably lightened the engine. But, nevertheless, the Type Test was as far away as ever and the specific fuel consumption was too high at 0.38 lb/bhp/hr, so I advised the Secretariat to cancel; yet nearly a year passed, after I had left the Ministry and returned to Associated Ethyl, before the Nomad was cancelled!

Some of the time before I was demobilized in 1946 was spent in reducing the size of the department to an assumed peacetime level, and I often recall the visit of a mixed American group of U.S.A.F. officers and civil servants from the Pentagon. They asked what the combined strength (numbers) of our procurement (production) and technical personnel had been in the War. When I gave a figure of about 400, distributed between headquarters, the outstations and firms, I remember how taken aback they were. Their figure was some eight or ten times ours.

Before the War ended, we had been seeing a good deal of Theodor Von Karman, the doyen of aeronautical scientists, who with his able team was touring around and getting acquainted with gas turbine progress, from us and also those Germans under interrogation. He was, in fact, enquiring into every form of scientific development stemming from the War.

I always liked Von Karman's way of explaining quite obscure (to me) scientific matters in very simple language. My wife and I became close friends with him and his diminutive sister, Pipo, who often accompanied him, particularly on official/social occasions. 'Pipo' died in 1951 and Theodor was greatly upset by her passing, which I always felt hastened his own end in 1963. Von Karman created the A.G.A.R.D. organization (Advisory Group for Aerospace, Research and Development) which was (is) allied to N.A.T.O. (North Atlantic Treaty Organization) and was carried on immediately after his death by his capable deputy Frank Wattendorf.

I left the Ministry and the Service after nearly seven years 'hard labour', and with so little leave that I could count the days away from work on the fingers of two hands, except for three weeks when stricken with infective hepatitis (jaundice) in the house of some good friends of mine, Bill and Audrey Martinhurst. Bill was then the Technical Director of British Thermostats and was concerned with the temperature and switching controls for engines such as the Merlin. I was dining with them one evening at Walton-on-Thames, not feeling too good, but the next morning I couldn't get up and Audrey made me stay in bed. Little did they know, poor things, what they were in for! I was, however, well looked after and have been forever grateful.

I think the Engine R. & D. Directorate in this period left a good legacy of engines that have proved successful and allowed the U.K. to maintain, for a time, the early lead given by Whittle, while others elsewhere were still at the learning stage. These engines, in alphabetical order: the Avon, Dart, Derwent, Ghost, Goblin, Mamba, Nene, Proteus, Python, Sapphire, Tay and, later, the Conway and Olympus, all came from the brains of our designers and manufacturers. We monitored and, I hope, encouraged their progress.

[1] *Development of Aircraft Engines and Fuels* by Robert Schlaifer and S. D. Heron. Published by Harvard Business School, 1950, and since reprinted.

[2] *British Aircraft Gas Turbines*, 9th Wright Brothers lecture, by H. Roxbee Cox, before the I.Ac.S. in Washington D.C., December 17th, 1945.

20   *Return to Civilization*

Upon returning to Associated Ethyl (as it still was) I started to renew our contacts with the motor industry, with Charles Good-acre assisting. We had engaged him just before the war from Austin's where, as an apprentice rising to an experimental engineer, he became a capable racing driver in their team of supercharged 750 cc cars designed by the late and ever-to-be lamented Murray Jamieson. Charles had also joined the Royal Air Force as an engineer officer at the beginning of the War.

One early and urgent matter demanding attention was to consider our future activities, in view of the fact that we had become a British company with some of the major oil companies as our shareholders. We were thus parted from the American Ethyl Corporation and its research facilities which we had enjoyed for some years. This led to an early decision to build an engine and fuel test laboratory, which the motor industry could come to regard as a source of help in assessing engine and fuel behaviour. Further, the laboratory would be sited within easy reach of Coventry and London. Eventually, Bletchley, on the A5 road and roughly half way between the two cities, was chosen.

One problem was that, while the laboratory was being planned and built, this would stay, or limit, our usual visits to the awaking motor industries of the U.K. and Europe, and also to the U.S. where we maintained useful contacts. So, after a discussion with Ray Bevan, I suggested we take on someone to work with Charles Goodacre, to supervise the planning, building and equipping of the laboratory and who could take on its management and running when built. I named a man I knew, Victor Lamarque, who was with M.I.R.A. (Motor Industry Research Association). Vic joined us in January, 1951, and we found him very conscientious and painstaking and most loyal and dependable. He managed the laboratory and largely contributed to its success, from its opening in 1953 until one day before his 65th birthday when, sadly, he died from a heart attack. His place was taken by Peter Dartnell, the present manager.

In this early post-war period, with frequent visits to the U.K. and European automobile industries, and trips to the U.S., we continued to meet our commitments and at the same time help our French aviation friends with plans to rebuild their engine industry.

Early on I visited Volkswagen at Wolfsburg, with Richard Morgenthaler, soon after it had been turned over to German control by the British authorities in 1949. Richard had, throughout the war and in the devastation of his own country, kept a close account of the tetraethyl lead situation, its manufacture, storage and secreting. He was one of the most dependable people I have known.

We were able to help Volkswagen with an exhaust valve problem in their early production 'Beetle', where exhaust valve life was then only some 6/8,000 kms (3,700–5,000 miles). By modifying the design somewhat and getting Teves, the German valve manufacturers, to use an improved steel, the valve life later went to 80,000 kms (50,000 miles). In this post-war period, with leaded petrol more generally distributed, reinforced by the experience of 80 octane leaded fuel that was universally supplied to the armed forces when America came into the war, we were better able to convince the motor manufacturers that we knew what we were talking about.

After the Liberation and before the U.S. visit I was approached, first in London, by French friends for advice on the reorganizing of their engine R. & D. etc. and also for views on possible future commercial aircraft to build; a matter on which I was only qualified to give a purely personal opinion. Therefore, when I returned from the American visit I went to Paris and talked with Air Ministry and Service Technique people, who arranged for me to see the airframe and engine industry.

I made a grand tour of the French Aircraft Industry with Colonel Guy du Merle (now Ing. General du Merle, ret'd) of the Service Techrique. General Sufferin-Herbert, a French Martinique and, like du Merle, a very charming man, was then in charge of the Service Technique. My good friend, Michael Golovine, who had served in the Air Intelligence of the R.A.F. during the war, accompanied us.

Most of the aircraft firms had their individual plans and there were no fewer than one hundred and seventy preliminary or possible projects. This was far too many for practical consideration or financing but, to be fair, the work was largely investigatory since the whole industry was restarting and seeking direction. One of our visits was to Leduc, the brilliant inventive genius who had designed the first example of a high speed aircraft propelled by an athodyd, i.e. an aerothermodynamic duct, meaning a tube fed by ram air from the speed of the aircraft, into which fuel was sprayed, burning the duct air which was ejected at high velocity to give jet propulsion. The duct or ramjet in this case comprised the whole fuselage length of the aircraft. This device was largely conceived and partly built during the Occupation, but successfully hidden from the Germans. Naturally, such a machine would first have to be accelerated before the duct could operate, either by being taken up under a normal aircraft and released or by fitting a gas turbine engine for the purpose. It was, in fact, tested in the former way. Once the fuel was ignited in the athodyd it would continue to burn.

Leduc was inclined personally to attend to each and every detail of his projects which could, and did, delay things. For instance, he had designed an ingenious, compact, integral gas turbine/fuel pump to supply the duct with fuel at high rates. This took considerable personal effort on his part and held up the main development.

Having completed this round of visits and given what advices I could, I was asked by Sufferin-Herbert and du Merle, on behalf of the Minister, if I would undertake an investigation of the engine situation and suggest how to get things moving and what engines to plan for the future. They desired that I concentrate principally on S.N.E.C.M.A., the national engine company, formerly Gnôme-Rhône.

I had already visited Turbomeca at Bordes, near Pau in the Basses Pyrenees. This Company, controlled by its presiding genius, Joseph Szydlowski, specialized in small gas turbines. On the occasion of my visit it was swarming with German engineers mostly from Daimler-Benz, headed by Nahlinger, their technical director. They were designing propeller turbines of 2,500 eshp and had brought with them the designs and mock-up of their (then) enormous turbojet of some 15,000–20,000 lb/thrust. My thought was that most of the Germans would eventually wish to return home, that the relatively large engines they were designing would have no immediate application (i.e. suitable airframe); that Szydlowski's inclinations were towards small engines and it would be far better to let him continue in that direction and support him. In the event Nahlinger returned to Daimler-Benz with some of his colleagues. One or two Germans elected to remain with Turbomeca, eventually taking French citizenship – one being Syring, the chief engineer.

I returned to Paris for the investigation of S.N.E.C.M.A. First, I thought I should have a French officer of high repute and integrity to accompany me, so I asked for Engineer General Martinot Lagarde. Martinot, a charming man, was not young since he was in his prime in the First World War. In fact, he went to Spain in 1915 to look at the 150 bhp, water-cooled, V8, Hispano aero engine, designed by Marc Birkigt in Barcelona. In 1916 he had brought it back to Paris to be built by the French Hispano company; it was afterwards manufactured by Wolseley in the U.K. and Wright-Martin in the U.S.

On our first official visit to S.N.E.C.M.A. in Boulevard Keller-mann, l'Humanité, the Communist paper, had banner headlines, somewhat as follows: 'Are the British to run (control) France. Montgomery at Fontainebleau and Air Commodore Banks to *close* S.N.E.C.M.A.' Field Marshal Montgomery was then Chairman of the Western Europe Commanders-in-Chiefs' Committee at Fon-tainebleau.

At that time, in 1947–48, S.N.E.C.M.A. had about twice the number of personnel they could usefully employ. Luckily, the French wanted all the skilled people they could get in the rebuilding of their industries and so numbers were eventually released for other work, leaving a total of about 3,500. At the Liberation there had been appointed as Minister of Defence for the Army and Air, a Communist by name of Tillon, and he appointed another Commun-ist as President of S.N.E.C.M.A., Weill. But by the time I went there both were gone and there was an 'administrateur delegué' at Kellermann but no President. The new Minister was a pleasant Burgundian, Maroselli, who had approved the Service Technique's suggestion that I be brought in.

S.N.E.C.M.A. was engaged on the continuing development of radial, air-cooled, piston engines, but also had a gas turbine in development designed by a German team from B.M.W. headed by Hermann Österich as chief turbine engineer. On the piston engine side, their pre-war types were being kept going for the various older aircraft and for spares and maintenance. There was, however, a large, multi-row, engine with, if I remember correctly, four rows each of seven cylinders. These cylinders were of forged light alloy with inserted steel liners, and I was rather disturbed when I saw the single cylinder unit then being prepared for its first run. Bearing in mind that light alloy has about one third the tensile strength of steel, and this cylinder was little thicker than a steel counterpart, it didn't take much imagination to see what would happen when it was subjected to explosion loads, plus the contributing bending stress imposed by the valve push rods. I was asked by the S.N.E.C.M.A. engineers what I thought of the cylinder design and I had to reply that I didn't like it and expected it to fail. It did and the cylinder parted company from the crankcase, I think on the first run. In the case of this particular engine, I advised that it be discontinued since a propeller turbine would come along and kill it. After all, we had already effected a similar exercise at Bristol in 1944; and Marchal, S.N.E.C.M.A.'s research director, was very receptive to reasonable suggestions.

The gas turbine was another matter. The German Bramo-B.M.W.

team, under the direction of Österich, was located at Decize, some 300 km from Paris, Österich being mostly on the spot in Paris. The turbojet, the Atar, was a simple, single shaft, axial compressor turbojet having an annular combustion chamber. It was an admirable engine on which the firm could 'cut its teeth'. Having studied its mechanical design and judging that the German team knew what they were talking about, I recommended that its development be continued and, if possible, accelerated. So as not to appear a great mogul, I got Stanley Hooker over to look at the design and give his opinion. He agreed that the Atar was a worthwhile project. They had also started on a larger engine, the Vulcan, but I thought the Atar would eventually achieve the Vulcan's thrust of some 8,000 lb. In any case one engine was sufficient for them to concentrate upon to expedite development.

During this period, shuttling back and forth between London and Paris, I was taken by du Merle to visit the Arsenal at Chatillon where they were developing a large vertical H-type, 24 cylinder, liquid-cooled, double crankshaft geared engine, based on four German Junkers engine cylinder blocks. The Arsenal engine had been breaking its crankshaft and they could not understand why. Looking at a broken crankshaft I could see that it was a good example of torsional failure. Then the Arsenal engineers startled me by saying that the crankcase might be too rigid and they were having one designed to give a greater degree of flexibility, 'to accommodate the crankshaft better'! I said that I had been criticizing (weak) crankcase design all my engineering life, yet here was a crankcase deliberately designed to give even less support to the crankshaft which would probably break more quickly if the crankcase didn't fracture first. In the event, it did.

I decided that a French chief engineer must obviously be appointed at S.N.E.C.M.A. with Hermann Österich as head of (Atar) turbine engineering. The man chosen was Michel Garnier. This choice was a good one and Garnier occupied the position for many years, raising the competence and status of the Company and its design and engineering capability. S.N.E.C.M.A. under Ing. General Ravaud is today one of the leading aero gas turbine manufacturers and its plant and equipment are first class. They have concentrated mainly on military engines, but this is natural since there was in post-war France, and in Europe, an earlier and greater demand for fighters than for commercial aircraft. But, with their links between Pratt & Whitney and G.E., S.N.E.C.M.A. are well placed to build a commercial engine as requirements come along. Various other interdepartmental appointments and changes were made, with Government (Service Technique) approval, and S.N.E.C.M.A. then advanced into the 'fifties under the control of a new President, Henri Désbruères.

Finally, in 1948, I was able to make an official report, with our recommendations, to Minister Maroselli and the Service Technique, signed by myself and Martinot Lagarde. This was accepted.

Almost at the same time, I was invited by A.F.I.T.A. (the French equivalent of the Royal Aeronautical Society) to give the first, inaugural, Louis Blériot lecture in May of that year. I chose for my subject and title, *The Art of the Aviation Engine*. It was delivered before a distinguished audience, including Theodor von Karman, on the 12th May. And although I had had the script translated by Michael Golovine and vetted by three of my French friends, Louis Armandias, René Lucien and Guy du Merle, I asked Guy if he would present it, since I did not think my French accent would last out the forty minutes or so of reading. Even if one is fluent in French, and I certainly am not, it is quite useless to expect a French audience to understand what you are saying unless you speak with the proper accent.

I was honoured further after the lecture by a vote of thanks from the doyen of French aircraft constructors, Monsieur Louis Bréguet. Others at the lecture included Roxbee Cox (Lord Kings Norton), as President of the Royal Aeronautical Society; Sir Frederick Handley-Page; M. Désbruères; Air Vice-Marshal Robert 'Bobbie' George, our Air Attaché; M. Jean Blériot, (son of his famous father); M. Potez; M. Roos; General Ziegler and M. Dassault.

As an aside, it was at Bobbie George's beautiful house in Paris that he asked me, at a cocktail party, to look after Greta Garbo. Far from being silent, I found her lively and entertaining and interested in quite everyday things.

After the lecture there was a cocktail party, followed by a banquet. Con, Ray Bevan, Michael Golovine, Louis Armandias and René Lucien were all there, supporting me. On the top table was the amiable and lively President of A.F.I.T.A., Jules Jarry, with Madame Blériot (Louis Blériot's widow) on his left. On her left was the Minister, Monsieur Maroselli (whose full title, if I have it right, was Secretary of State of the Forces, Army and Air), then myself. There were many other high Air Force officers, civil servants and chiefs of the French and U.K. aircraft industries. At the end of the banquet, Jarry made his usual fluent and witty speech, some of it in English, with a toast to myself. Then Maroselli rose and made a graceful speech on, 'the help I had given France and her Aircraft Industry'. This was really in the form of a citation since, at the end, he invested me with the insignia of a Commander of the Legion of Honour and kissed cheeks with me. I felt very proud at that moment and gave a short speech of thanks, in bad French, for the honour he had done me. Jarry then presented me with a medal he had had specially struck by the French mint to mark the occasion. It is a typically good example of French art, showing the cliffs of Baraques and Dover and Blériot's machine in the flight across. He also had replicas for the specially invited and distinguished guests.

Our company work continued in the U.K. and Europe, with fairly regular visits to the U.S. Mostly, we concentrated upon the automobile manufacturing countries, France, Germany, Italy and Sweden. In France it was interesting, in the post-war period, that Citroen continued to maintain their reputation for aloofness though they, too, had valve troubles. We usually met the engineers concerned in a sort of office-cum-waiting room; never at that time were we invited inside the plant proper. This was always particularly exasperating to us and our French colleagues. Renault were different and let you in to see things, though they could be almost as difficult as Citroen since they were not convinced, even then, that leaded fuel had come to stay. But, eventually, we established a good rapport with their engineering staff, headed by Picard, and were able to help materially in dealing with their valve problems once they took notice of us. On the other hand our relations with Fiat had always been good, going back to the early 'thirties, and we made

The Author after being appointed a Commander of the Legion of Honour in Paris, May 1948.

frequent visits to Turin to discuss automobile and aviation matters with Dr Carlo Bona and Professor Guiseppe Gabrielli, head of the Fiat aviation department.

In my continuing liaison with the French aviation authorities, I met, towards the end of 1949, the new Minister of National Defence, Monsieur Pleven. He had in his earlier days been with a firm in Liverpool and spoke excellent English, which was a relief because my French was (is) so bad. Also, I was to meet the new French Chief of Air Staff, General Léchères and his very bright Chief of Staff, Colonel Pierre Gallois (later General Gallois and now with Dassault). Léchères and I became good friends and we both trusted each other. Léchères, who followed the distinguished General Vallin, had great commonsense backed by experience and foresight. He and Pierre Gallois were complementary but quite different in personality, Léchères being quiet in manner and Pierre ebullient with a needle-sharp mind. I had regular meetings with Léchères and we mostly discussed future trends in turbojets and fighters and the progress of S.N.E.C.M.A. with its new management under the gentlemanly Henri Désbruères. As a result of my investigation of

General Bernasconi, Commander of the Italian High Speed Flight, standing before the Memorial to the Flight when unveiled at their base at Desenzano, Lake Garda, 1966.
Photograph: General Mario Bernasconi.

S.N.E.C.M.A. we agreed with the Service Technique that the firm should continue along its own lines to progress in turbine technology using the Atar as its basis. We had also advised that Hispano-Suiza build a Rolls-Royce turbojet under licence. This eventually happened and they took a licence for the Nene 2, a centrifugal turbojet of some 5,000 lb (2,273 kg) static thrust. Later, they negotiated the Tay licence from Rolls. Of somewhat greater thrust than the Nene, it was developed to give 7,700 lb thrust (3,500 kg) and named Verdon. Then the Bréguet-designed Atlantique, a maritime reconnaissance aircraft, was fitted with the Rolls-Royce Tyne built by Hispano under licence. This is now a truly international engine with components manufactured in Germany and other countries. In this post-war period, Prince Poniatowsky had handed over managing directorship of Hispano to Maurice Heurteux. 'Ponia' had come to the U.K. before the War, or in the early war period, to organize and run the (Hispano) firm at Grantham for the manufacture of Marc Birkigt's 20 m/m gun (canon).

In the middle of 1952, Ray Bevan was approached by Air Marshal Sir John Boothman, Controller of Research and Development (C.R.D.) of the Ministry of Supply, who asked him to loan me for about a year to run my old engine Directorate. Luby, who had been my deputy and become my successor, was leaving to join Rotax (now Lucas Aerospace). I must say that I thought Ray very patient and understanding. Whatever he may have thought, I was not all that keen myself to go back, after having served there for so long. It was, however, agreed that I should go for a year, as 'a dollar a year man' – an American term used when the U.S. Government paid professional people on loan from their companies or institutions a token of one dollar per annum. Associated Octel (ex Ethyl) continued to pay my salary during the loan period. I was joined by R. H. 'Bob' Weir as my deputy. He was a specialist in gas turbine performance and of great assistance to us.

It so happened that I had been asked a few months previously to give a lecture to N.A.C.A. (now N.A.S.A.), in Cleveland, Ohio, on British aero engine procurement procedures. I had already submitted the manuscript to M.O.S. for formal clearance, though it contained no secret or restricted information. Therefore, I was very irritated, when already in my Principal Director's job, to be told that the Minister, Duncan Sandys, did not wish me to give it. Obviously, at that late stage (September) with the lecture due in November, I could not contemplate letting down N.A.C.A., so I told the Permanent Secretary that I would leave the Ministry and give the lecture rather than risk a showdown with my American friends. This seemed to have some effect and, eventually, I was told that the Minister had withdrawn his objections.

I have already mentioned that my first problem on returning to the Ministry concerned the Nomad. With the cancellation of the Naiad, Napier had already started on the design of another, larger, propeller turbine, the 3,000 bhp Eland. In cancelling the Naiad an R. & D. contract was given Napier for continuing work on the compressor, which was fairly efficient.

The Eland was to be largely a private venture (P.V.), helped by Ministry R. & D. funds for the (Naiad) compressor work. Therefore, I thought it necessary to warn Sammons, the managing director, that the project would be very costly. But 'Sammy', ever the optimist, had apparently convinced his Board that it would all be worthwhile. Perhaps we should have been firmer at the Ministry in discouraging it. At the time of the development of the Napier Gazelle, their helicopter engine, I felt strongly that we had too many engine manufacturers in the country for the useful contracts we could give, spreading a limited amount of butter very thinly and satisfying few. So, I had a talk with Hives (Lord Hives) and asked him if he would consider approaching English Electric to take over Napier. He first spoke with Lazard's, the bankers, who made a tentative approach to the Company but with no result.

Unlike the Nomad, which was intended for a maritime reconnaissance aircraft, the Eland had no foreseeable application. A year or two later, however, when the engine had reached the civil certification stage, Napier managed to persuade Convair, manufacturers of the CV240 twin engine airliner (also the 340 and 440), designed originally to take the place of the DC3, that the 'Eland' would make a good replacement for the Pratt & Whitney piston engine. A trial installation was made on a CV340 (or 440) by Pacific Aeromotive Engineering Corporation, but further installations for the airlines proved uneconomic and the scheme was abandoned. Napier had to bear the cost of the re-conversion back to the P. & W. engines which, taking the whole development cost of the Eland, must have been pretty expensive for them.

There was, however, another Convair conversion (named the CV580) with the Allison 501-D13 turboprop in the CV340 (or 440). This was F.A.A. certificated in 1964 and a number were operated by Frontier Airlines and, later, Allegheny. Some are still in operation.

The lesson to be learned from the Convair/Napier exercise is that it is always costly to have relatively few engines far from home and factory, where the effort to keep them going involves much time and travelling for a number of technicians and field service personnel. Also, a successful aero engine can now remain in service for anything up to thirty years and the manufacturer will be involved for a large part of that time since the aircraft licensing authorities can demand the immediate incorporation of a modification, placing the onus on the manufacturer and making him legally responsible. Naturally, after an accident a subsequent enquiry into airworthiness could cast doubts on the engine's modification standard.

Within a few weeks of my rejoining the Ministry, all departments were told that they had to reduce their R. & D. budgets by 10 per cent. This across the board exercise always seemed stupid to me and could be damaging to progress where some departments might need more funds while others could cut by more than the stipulated 10 per cent. The first notification we had of this was on a Thursday prior to a meeting on Saturday morning. The meeting was under the Chairmanship of (the late) Scott Hall, Director General of Technical Development (Air), who said that he would take the Engine Directorate's case last, since we had 'the largest and most complex budget'. The meeting droned on for most of the morning, with one or two members impatient to get away for the weekend. Generally, they accepted the imposed cut without protest, yet I knew some could have taken reduction of 15 or even 20 per cent. It was about 12.35 p.m. before 'Scottie' came to me and I had to say that, in view of the new engine projects in hand, and particularly the new Mark of Avon, I couldn't cut but would probably want more money later. This caused a furore and the Deputy Secretary, Finance, said that I couldn't ignore the instruction. I said that I could if I were to remain responsible for those engines in development and expected to be ready for their aircraft i.e. the Proteus for the Britannia and Olympus for the Vulcan bomber, in addition to the Avon RA14. Poor Scott Hall, a good friend of mine over some years, held his head in his hands, not knowing what to do. I challenged the Deputy Secretary to make the engine cut himself and take the consequences. This he wouldn't do. Eventually, Scott Hall said that he and I and the Deputy Secretary would take it up with the Controller (Boothman) on the Monday. This was done and I got no cut.

There are two things here: first, I was only a temporary civil servant and if pushed too far could resign and go back to my firm. Such a stand would not be easy for a permanent civil servant, with his career to consider. Second, it shows how almost intransigent is the Treasury through its appointees since there was no engine at that time without a specific aircraft requirement, except possibly the Nomad, and even that had been considered by Avro for a maritime reconnaissance aircraft they were scheming. To risk slowing everything down by denying funds could have seriously affected aircraft deliveries, particularly the Britannia for B.O.A.C.; and it would cost even more in the long run to re-start or accelerate the project were it to be run down. Further, an airline can ill afford to wait for a new aircraft when its introductory phase and operational rôle have already been planned.

On the other hand there have been occasions, when it was 'politically convenient' that money be made available at the drop of a hat. I remember Roy Fedden getting a verbal O.K. from Sir Stafford Cripps, (at the time Minister of Supply), that he, Fedden, could tender for a development contract for a small propeller turbine he wanted to build. Not long afterwards Cripps went as President of the Board of Trade (after Labour came to power in 1945) and a new Minister, John Wilmot, was appointed to M.O.S. Wilmot was immediately approached by Fedden, who told him of the Cripps promise, but since I was then trying to limit the number of post-war engine manufacturers, I did not wish Roy to be further encouraged. However, he went to Cripps, who confirmed to Wilmot his undertaking that Fedden could tender for his (Cotswold) propeller turbine and I was asked to see to it. In due course, we received a tender for the design and development of the Cotswold with the supply of six engines. Remembering that Fedden had to get a factory and assemble a design staff, a month or two in submitting the tender was pretty quick going. But he, in his usual way, had gone ahead on the original Cripps promise, so I suppose his early action wasn't too surprising. However, the estimate of costs he had submitted was much too low to be realistic. We had already ordered the Dart and the Mamba and consequently knew what established firms (and our own experience) considered feasible. While I did not in the least blame Fedden for trying, we could see that it was going to be almost impossible later to justify the extra funding which would inevitably be needed in order to give continued support for the project.

I stuck out against accepting the figures or giving a contract, but yet another visit by Fedden to Cripps at the Board of Trade resulted in a meeting being called at the Chancellor of the Exchequer's house, No. 11 Downing Street, to which I went with our Controller, Edwin Plowden (now Lord Plowden). Cripps was there and I was asked to give him and the Chancellor (Dr Dalton) my views. Afterwards, Cripps turned to Dalton and said something like this, 'I think, Hugh, we can start Fedden off with what he wants' – and Dalton agreed! So, that was that. I was demobilized a year later (1946) and had left the Ministry when the funds dried up and poor Roy *did* want more money, much more. He had also designed and built a prototype, rear-engined, small car having a three cylinder, sleeve valve, 'Y' type air-cooled engine. That also had to be abandoned. I was truly sorry for him, but he did not realize that conditions were not the same as those after World War I, when he was a pioneer. Now his old firm (Bristol) and Rolls-Royce were already established in size and capability (know-how) and he was seeking to start again in a new technology more than twenty years later.

When I took up my position in the Ministry in 1952, I was early assailed by the National Gas Turbine Establishment and others on the 'square/cube law'. This would, they averred, act against increasing the size (thrust) of a turbojet above about 15,000 lb thrust, some even suggesting a limit as low as 10,000 lb. In simple language, this law means that, all other things remaining the same, increasing the engine size above a certain optimum would result in increased scantlings (larger – heavier – measured size) with consequent increase in specific engine weight, i.e. more pounds weight per pound of thrust. But this argument usually ignores the advances in the 'state of the art' that have come as a result of the development and subsequent operation of preceding engines. Then there are improvements in materials, manufacture and component efficiency. Therefore, I wasn't being too prophetic when I told my colleagues (in 1952) that I saw no reason why engines of 50,000 lb/t would not be coming along in, say, twenty years' time and 100,000 lb/t engines would be flying aircraft within the lifetime of many. The former prediction has already come about.

During my time at the Ministry I had had several talks with Sir Reginald Verdon-Smith, then Managing Director of the Bristol Aeroplane Company, who suggested that I might like to join them sometime. I was most interested, but since Associated Ethyl had made up my salary during the War I felt I should give them more time when I eventually returned.

V-S's suggestion was particularly attractive because the piston engine, on which I had worked for so long in aviation, had already been virtually swept away by the gas turbine and I wished to continue the work I had been doing with this new prime mover. In 1954, I was again approached by V-S and asked if I would take on the managing directorship of Bristol Engines, with a seat on the main Board. I then accepted, but when V-S put it to the main Board for formal agreement, Sir Stanley White, the Deputy Chairman, felt that Norman Rowbotham, the managing director I was to replace, should remain on the Board. V-S, quite fairly, put this to me by telephone immediately after the Board meeting. I hesitated and finally, regretfully, refused. Air Marshal Sir Alec Coryton was eventually appointed as Managing Director of Bristol Engines. Later in the year, I joined the main Bristol Board with a seat on the aircraft and engine companies' boards.

The Britannia was being built and flight tested when I arrived, but I was asked if I would visit North American Aviation in the U.S.A. and try to sell them the Olympus for their new fighter, the F100. An Australian mission was visiting Europe and the U.K., returning home through the U.S., and were expected to look at the

F100 for their Air Force. I arrived in Los Angeles with Basil Black-well, one of Bristol's engine performance specialists. We saw Lee Atwood and Ray Rice, President and Vice-President of Engineering respectively, also 'Dutch' Kindelberger, the Chairman of the Board.

The first difficulty arose because the Americans would not do their sums without considering an after-burning engine. So I had to ring Bristol and ask what diameter and length they would recommend for an afterburner. These dimensions were needed to determine if any alterations would be needed to the fuselage. I was told, 'If the Australians want the F100 with the Olympus, then Bristol would arrange to fit an afterburner'. I replied that that wasn't good enough and we had to give an answer then and there. The upshot of this was that we had to look around for an American afterburner. I saw Marquardt and Solar, both making good A-B equipment. Eventually, the Solar afterburner was chosen because we were offered a good deal by Ned Price, head of Solar, who was an ardent anglophile. Ned lived a good deal of the time in London after his retirement, but died a few years ago – a loss regretted by many here.

In the event, the Australians did not buy the F100 but they eventually took a manufacturing licence for the F86 and fitted the Rolls-Royce Avon engine. It was the North American built F86 that did so well in the Korean War against the Russian Mig.15. The Australians had a Gloster Meteor squadron (No. 77) in Korea, but the Meteor could not compete with the Mig and it was later confined to the ground attack rôle.

On my return I was asked by Verdon Smith, early in 1955, if I would look into the planning of the Britannia, five of which were due to be delivered to B.O.A.C. in the Autumn of that year. Since the C. of A. (Certificate of Airworthiness) was not due until August, it appeared unlikely that the five aircraft could be delivered on time. I started by asking the engine company for the loan of Warren Snell, whom I had known for some years and who was a very good 'ferret' and able to get the information we wanted. About 50 per cent of the Britannia was sub-contracted, so our first job was to find out whether the sub-contractors had drawings and the necessary jigs and tools. Warren did a first class job on this, though it took us nearly two months before I could report to the Board that we would get only two of the five aircraft promised at the end of the year–that was if the C. of A. came on time. Two were in fact delivered to B.O.A.C. in the last days of December 1955. We were helped by taking on a very energetic and capable Australian by the name of Don Stewart whom we made responsible for final preparation of the machines prior to flight testing.

Then there came all the misery of engine intake icing, which took a long time to cure and was one of the principal reasons for the delay in getting the Britannia into service, in fact some two years after the original and estimated delivery date. While the Proteus was basically a good engine it was designed with a back to front compressor, having the air inlet turned through 180° and located roughly in the middle of the engine. The arrangement was chosen because it was intended for the large Brabazon air frame where it had to fit between the main spar and the wing leading edge. We hadn't then conceived long engines, with their tail pipes projecting back to the wing trailing edge. The 180° air intake bends tended to ice up at or near 0°C in cloud and in high humidity. Eventually the icing problem was cured by fitting Napier electrically heated mats at the bends.

The Britannia proved to be a good aircraft and a very quiet one; it was only a pity that it was late and already overshadowed by the turbojet aircraft like the Comet and later the Boeing 707 and Douglas DC8.

I have always held that it is not good policy, for obvious competitive and business reasons, to combine in one firm the manufacture of aircraft with that of engines. Other aircraft manufacturers are not likely to buy your engines if you are competing with them in aircraft, and particularly if there is a suitable engine available elsewhere. This was brought home when I was asked to see George Edwards (Sir George Edwards) of Vickers, who was designing the VC10 (one of the nicest aircraft inside and the noisiest outside) to try to sell him the Olympus. His reply was to the effect that while Bristol built aircraft competitive to his he would go to Rolls for engines. He was also designing the Vanguard for B.E.A. and Air Canada and for this machine he took another Rolls engine, the Tyne turboprop, but I have always thought the Government should have stepped in and told B.E.A. to take the Britannia until a suitable turbojet came along. True, it would have lost a customer for Vickers but even with B.E.A. and Air Canada together these two airlines did not justify the Vanguard. The only British airliner that has enjoyed worthwhile production and sales since World War II is the Viscount with, in more recent years, the B.A.C.111 a moderately good second.

While at Bristol I was asked by Société Franco-Britannique, our French representatives, to meet Wibault, the aircraft designer and constructor. His was the originally thought out idea on vectored thrust and he had designed a flight vehicle which had four snail type centrifugal blowers down each side, the casings of which could be rotated or vectored. These eight blowers were to be driven through shafts and bevel gears by the Bristol Orion engine, a propeller turbine then under development designed by Stanley Hooker's team. I considered this original arrangement of Wibault's somewhat complex and mechanically cumbersome. Also, we thought at Bristol, that the Ministry of Supply would eventually cancel the Orion; so I asked Stanley Hooker if he would come over and meet Wibault and look at his scheme. This resulted in the classic Pegasus engine, based on the Orpheus, with integrated or integrally vectored thrust. It was *not* supported by our Government but was financed by the U.S. through the Mutual Weapons Development Procurement Office (M.W.D.P.) in Paris. The Paris office was then run by Colonel O'Driscoll, who had obtained funding for the development of the Orpheus powering the Fiat G91 light fighter for N.A.T.O. After his tour of duty he handed over to Colonel (later General) Chapman who arranged for 75 per cent funding for the development of the Pegasus engine, Bristol supplying the balance. The prototype Pegasus first ran in the autumn of 1959 and was fitted in the P1127, prototype of the Harrier, about a year later.

As a matter of fact it took some little time after the general acceptance of the turbojet as a fighter engine before we could get Hawker's designer, Sydney Camm (the late Sir Sydney Camm) interested in jet propulsion; so early in 1944 we, in the Ministry, with test pilots 'George' Bulman and Philip Lucas, got Hives to invite Sydney to Duffield Bank, the Rolls guest house, for the weekend. There, he met Dr A. A. Griffiths, the Rolls head of research and forward projects, Lovesey, the chief development engineer, and others. After that weekend Sydney was convinced about the turbojet and later schemed out the P1040, after getting details from Rolls of the B41 (the prototype Nene) engine in the autumn of that year. This was developed into the Naval prototype N 7/46, which in 1953 went into Fleet Air Arm Service as the Sea Hawk.

The tragedy of the DH121 or Trident is worth some comment since it could have been an outstanding success. It was conceived by de Havilland before their takeover by Hawker Siddeley. The first public announcement that a new short-haul aircraft was in the offing was made by B.E.A. early in 1958. Then, for no good reason that anyone could fathom, there was the formation of the Airco group to handle the manufacture of the Trident – composed of deH Fairey and Hunting. Airco was the revival of a name of a company formed by Holt Thomas in World War I, with Geoffrey de Havilland as its designer. deH left later to form the de Havilland Aircraft Company in 1920. Rationalization was the theme of the Tory Government in the late 'fifties and 'sixties, which put an end to Airco because Hunting and de H found themselves to be in different camps when the latter was taken over by Hawker Siddeley together with the Blackburn Group.

The original Trident was designed by the de H team for 111

passengers and to have the Rolls-Royce Medway engine of some 14,000 lb thrust. Classified the DH121, the Trident was about the same size as the Boeing 727. Then, apparent insanity hit B.E.A. who decided that the Trident must be reduced in size. They were even willing to pay de Havilland for the design and other work already done on the original de H 121. In speculative discussions on this peculiar move, that have gone on from time to time since the B.E.A. decision, the airline have always emphasised that the reduced size fitted *their* route pattern and it was they who were buying the aircraft. Nevertheless, it seemed very short sighted for a national airline not to consider foreign sales if only to help cut production costs and advance future development of the design. It was a thoroughly bad decision in that, since the Trident went into service in 1964, two larger models have been built which, even with the Chinese orders, only brings the total number of Tridents to just over one hundred – in over ten years! Yet the very similar, and probably no better, Boeing 727 has become the biggest selling jetliner in aviation history.

I don't suppose that the Trident would have achieved orders on the Boeing 727 scale, but in ten years there could have been several hundred or more. Beverley Shenstone, one time chief technician to B.O.A.C. and B.E.A., once said to me, 'You don't buy a new aircraft that is smaller than the one it is to replace'.

It could be asked: what about the Viscount? The answer is that the '630', the Viscount prototype, with its Dart engines, was designed and built at Government expense, ahead of any B.E.A. requirement, experience or thinking. It was Peter Masefield, as Chief Executive of B.E.A., who eventually took it on and had the passenger capacity increased from 32 to 47, i.e. the type '701'.

Until the Viscount, there were no basic transport (airline) aircraft of classic type (like the Douglas DC2 and 3) built in the U.K. before the War. There was the makeshift Vickers Viking, after the War, and the Airspeed (D.H.) Ambassador and the Avro York, though none could reasonably be regarded as examples of our future capability. The European aircraft industries had some almost classic types before the War, such as the three engine German Junkers Ju 52, certain Fokker machines and the French Wibault and Dewoitine aircraft. We had flying birdcages like the Armstrong Whitworth Argosy and Handley Page HP 42, very safe and very slow. I had quite often been landed in a '42' at Lympne to take on sufficient fuel to complete the flight (from Paris) to London against a head-wind – 90 mph was its normal cruising speed.

In more recent times there have been some encouraging and practical signs of European collaboration such as in the Concorde, the A300B Airbus and Jaguar and M.R.C.A. strike fighters. But apart from Concorde which, in my opinion, is a bold experiment, it is difficult to compete in commercial aircraft against the U.S. with its very large domestic market. Europe is not (yet) the 'United States of Europe' and until it is, (if ever), some countries just do not have the resources to support a worthwhile aircraft industry. Therefore, the aviation products of America cannot be ignored either from the military or commercial aspects. However, since the aircraft and space industries embrace practically all the highest and most advanced technologies and because a country not taking part in any aerospace activity will inevitably lag behind the leaders and lose out in the World's markets, it is necessary that those European governments who can should maintain their support in this field. But they will have to be critically selective in such support so as to ensure real advances in the state of the art, though the U.K., France and Germany could and should get together on such projects as the A300/B10. It is not enough that these talented aviation nations should be demoted to the rôle of sub-contractors to the U.S.A.

I am not so apprehensive of engines, either British or French, because there is already a genuine feeling by both countries of the need to co-operate with the Americans, if only to avoid unprofitable competition. This inevitably leads to such (US/European) collaboration, but there must be included in these arrangements design and development in addition to production participation if such deals are not to be regarded as little more than a super sub-contracting job.

I had been with Bristol for about four years when it was decided, between Verdon Smith and Roy Dobson (Sir Roy Dobson, Managing Director of the Hawker Siddeley Group), that Bristol Engines should join with Armstrong-Siddeley to form one large aero engine facility. Sir Arnold Hall, who was on the Hawker Siddeley Board, became Managing Director of the new Company, Bristol Siddeley Engines, and I was appointed Sales Director. The head office was in Mercury House, Knightsbridge. I did not like the idea of top management being remote from the main plant, in this case Bristol. Armstrong-Siddeley were in Coventry and, it was argued, London was central to the two; but it was so easy to get in a car at Bristol and motor to Coventry. My view was, and still is, that senior staff generally want to see the boss sometime during the day. In any event, we all had to go to London from time to time for visits to the Ministries and to meet foreign visitors.

Anyway, I did not feel happy so, quite amicably, I asked V-S (who was Chairman of the new company and also of the Bristol Aeroplane Company) if he would release me from my five-year contract. It had only a month or so to run. He gave me this release although, at the age of 61, I realized I was taking a big risk since I still wanted a job.

All this occurred in the late summer of 1959, at a time when there happened to be a journalists' strike on, so few knew that I had left. But, one day, as the aviation bush-telegraph tapped away, Eric Turner of the Blackburn Aircraft Group rang me and asked if it were true that I had left Bristol. If so, would I like to join him on the Blackburn Board. I said 'Yes' and joined Blackburn just before the Farnborough Air Display that year. I was based then at Blackburn's London office but went frequently to Brough as Eric wanted my views, engineering-wise, on various matters, mostly on the engine company. I should add that, just before the War ended, Robert Blackburn, the founder of the Company and a good friend over many years, asked me what I thought of Blackburn's taking up the manufacture of small gas turbine engines. I was against it and told him that it would absorb a lot of money to get started and keep going. He was enthusiastic and wanted to go ahead, so he did a deal with Joseph Szydlowski, 'le patron' of Turbomeca, which resulted in a licence agreement being negotiated for Blackburn to build his turbine engines. Blackburn were already in the small piston engine business and manufactured the Bombardier, an engine of their own design but on D.H. Gipsy lines. However, this Turbomeca deal was 'a horse of quite a different mule'. Blackburn's had then been divided up into three separate companies, aircraft, engines and electronics, and while this was quite a reasonable division there were only really sufficient funds, even with Government contracts, to keep the basic activity, military aircraft design and manufacture, going. When they took the Turbomeca licence the Blackburn people made the cardinal error of modifying the design of the engine; that infuriated Szydlowski who was not 'le patron' for nothing. Of course, some detail modifications were necessary to meet various aircraft and helicopter requirements, but Szydlowski hated *any* alteration. The real mistake was not to go and discuss it with him in the first place. Soon after joining the Blackburn Board I was asked to take over the engine company as managing director. This entailed travelling from London to Brough every Monday, returning home on Friday or Saturday.

The electronics company also grew, because Blackburns had some good electronics people working on flight test equipment. Various bright ideas developed which resulted in the manufacture of specialized equipment that became attractive to others who saw the advantages of buying more or less off the shelf. The difficulties began to arise when new ideas spilled out as equipment was further developed, or new products were thought of with a possible sales potential. All this, however, needed funds and Blackburn were just not wealthy enough to advance these and await a return, particularly while competing against the very large electronics concerns.

Joining Blackburn was pleasant because, until I went to Bristol, I had always been with small firms, which were more intimate and had a lower moment of inertia where quicker decision making was concerned. But I hadn't been with Blackburn's for more than about three months when we were taken over by Hawker Siddeley, and so I was back with the biggest aspidistra of them all. Eric Turner had already left and gone as Chairman of B.S.A., his place being taken by Arthur Jopling. My earlier prayers were, however, answered inasmuch as the engine company was absorbed, with de Havilland Engines, into Bristol Siddeley Engines.

When I was running Blackburn Engines I stayed at the Station Hotel in Hull, a nice place but a bit noisy if one's bedroom was on the station side. Early one morning, at about 3 a.m., there was a locomotive standing in the station blowing off steam about every few minutes or so, and I could not get back to sleep. When I went to breakfast I mentioned the disturbance to the Hall Porter. He said: 'You know, Sir, the same thing happened to Sir Donald Wolfit (the Shakespearian actor) when he was playing here. He rang me from his bedroom and said, 'Portah, Portah, can you tell me when this bloody hotel gets to Kings Cross'.'

I was later appointed Chairman of A.T.S. (Aerospace Technical Services), a small forward thinking technical group within H.S.A., with my friend (the late) Michael Golovine as managing director. This was a most interesting assignment and the work varied from giving John Lidbury (now Sir John, and at the time managing director of Hawker Siddeley Aviation), a true estimate of the likely cost of the HS125 to collaboration with European scientific bodies and firms on aerospace matters.

One day in 1963 John Lidbury 'phoned me to come and see him; he then asked me if I would go to the de Havilland Division at Hatfield and take charge for a period. Since I was already in my 66th year and had thought they might consider retiring me, I was very pleased to take on the Hatfield assignment. When I joined them they were engaged in preparing the Trident and the HS125 executive jet. I knew something of the '125' because Golovine and I, at Lidbury's request, had given him quite an accurate estimate of its probable cost, which was between two and three times the original estimate.

My first week at Hatfield was largely spent in going round the plant with Gilbertson, the Works Manager, and attending two technical meetings which Wilkins, the Technical Director, had called – one on the Trident and the other on the HS125. I remember the latter meeting, which was to determine what could be done to reduce the 125's production cost, already badly overrunning the original estimate. But when suggestions were put to the meeting that the windscreen wiper at the right hand pilot's position should be eliminated and the 'punka-louvres' also done away with, I had to chip in with the remark that this was impracticable. The '125' was virtually an airline machine, operating at the same altitudes as jet airliners, therefore it would have two pilots and they had to be provided with duplicated equipment for flying under all conditions. It was not, I said, a '450 mph Dove'.

That week, John Cunningham, our chief test pilot, took me for a flight in the '125'. I was in the second pilot's seat, just to get the feel of the aircraft. We flew through the edge of a thunder-storm with some hail and I heard a sharp crack from the cockpit roof. When we landed, it was discovered that some of the balsa wood fairing around the roof had peeled off and disappeared. This was the first time that I knew of the balsa wood fairing and during my four years at Hatfield I tried to get rid of it and have a deep pressing for the cockpit canopy. This was, apparently, a difficult and costly modification but, despite delays in getting the '125' certificated and into production the modification was eventually done.

A delicate matter came up when the question arose of the likely number of hours needed for the flight certification of the Trident I. A certain number of flight hours had been allowed for, based on Comet experience, but A.R.B. thought them insufficient. Our people didn't wish to go beyond the original figure because of the extra cost and so there was something of an impasse. We eventually had a meeting of all concerned in my office, with Bob Hardingham (now Sir Robert Hardingham), Chief Executive of the A.R.B., and David Davies, the A.R.B. chief test pilot, present. I had previously mentioned to my colleagues that the Comet certification was some years back and A.R.B., through Davies, had acquired much more experience and cleared later aircraft such as the Boeing 707 for B.O.A.C. The meeting was amicable and agreement reached on, if I remember rightly, some 200 extra flight hours.

I had my 1963 Christmas in Japan. We were asked to go there at short notice since All Nippon Airlines were interested in a stretched version of the Trident which, although it would have involved a major re-design, nevertheless offered a chance of the additional sales we so badly needed beyond the B.E.A. order. This version would have been somewhat larger than the Trident I, but could only offer an increase in passenger capacity at the expense of range since the Spey engine, which had replaced the Medway, did not have the extra thrust needed.

After a number of visits with our performance and manufacturing colleagues, we lost to the Boeing 727. It was all rather sad, but we managed shortly afterwards to get an order for three Tridents from

Pakistan Airlines in Karachi. This meant a lot of hard work for Cecil White (Sales Director) and our team, and a number of visits before the contract was signed; and a lot more work when the machines were introduced into service. Unfortunately, we had accepted an unrealistic air conditioning requirement from the airline for a Freon system demanding a cabin stablization temperature down to 60°F in 30 minutes from a high ambient temperature of 120°F. But passengers entering the cabin from such a high outside temperature would have suffered considerable shock. It was eventually agreed to stabilize the cabin temperature at 75°F in the 30 minute period and this figure was quoted as an estimated performance from a relative humidity of 40 per cent.

The object lesson here is that it is contractually risky to undertake a completely new requirement without any previous technical background. The financial penalty for non-compliance can be, and was in this case, high. However on the credit side the HS125 was beginning to sell well, mostly in America, from the mid-sixties onwards.

After being four years at Hatfield, four good years with very pleasant colleagues, John Lidbury wished me to come to head office at Kingston-upon-Thames for my last year before I retired at the age of 70. My time was mostly spent in travelling around the World, visiting customers, agents and distributors. Air Chief Marshal Sir Harry Broadhurst took over civil aircraft, running it mostly from Kingston with Jim Thorne, the General Manager and Director, remaining at Hatfield.

I can only say here that J. P. Smith and Peter Owen and their colleagues have done a good job of designing and engineering the Trident which, being a considerable and understandable advance on the Comet, has demanded much greater effort and attention to detail in every respect. We took on Johnnie Johnstone from East African Airways, knowing his reputation there for maintaining their Comets in service. He reorganized our Product Support and Spares set-up to some effect. Alf Sewart (and later George Wilkinson from my old firm, Blackburn's) was in charge of production while Jack Garston and his colleagues and staff ran the Chester plant and were responsible for launching the HS125 into production. All these and many others with them have made the Trident and HS125 good and reputable aircraft.

Coming back from a round trip that had taken me East, in stages, to Australia and New Zealand, returning West about, via the U.S. and Canada, I made a routine visit to our distributors, AiResearch Aviation Services (Garrett-AiResearch) at Los Angeles, to see how the sales of the 125 were faring. This was my last trip for Hawker Siddeley Aviation before I retired at the end of March. While at AiResearch Aviation Services I got a 'phone call from Walter Ramsaur, then Senior Executive Vice-President of Garrett, asking me to come and see him. During our conversation on this and that, he asked me if I would be interested in becoming a consultant to Garrett when I finally retired. I told him that I would be delighted. Walt was as good as his word, and more or less immediately after leaving H.S. I became consultant to Garrett; and from Harry Wetzel, their President, across the board at all levels in the Corporation I have been treated with great consideration.

At Garrett, I met top level German engineers. One, Dr 'Ted' Von der Nuell, I had seen on several occasions when visiting D.V.L. (Deutsche Versuchsanstalt für Luftfahrt, Berlin-Aldershof). His work as a specialist on centrifugal superchargers and turbo-blowers for piston engines was outstanding. The other engineer, Helmut Schelp, whom I had met at the War's end when he was under interrogation by us, specialised on gas turbines and was responsible, in the German Air Ministry, for encouraging the engine manufacturers to move into the gas turbine field and embark upon turbojet development. In this he was most successful. These two, when later joining Garrett, were, with Homer Wood, largely responsible for launching that firm's gas turbine programme for A.P.U.s (auxiliary power units), small propeller turbines and fan-jets for executive and general purpose aircraft – also industrial turbo-blowers for diesel and gasoline engines.

In 1967, and up to my retirement, I had taken part in four S.B.A.C. missions, two to Japan, one to Australia and then an official mission to Moscow for the massive air show put up by the Russians to celebrate the 50th Anniversary of the Revolution. The Show was actually held some months ahead of that date.

The Moscow Air Show mission included Adrian Lombard, the Technical Director of Rolls-Royce. I suggested to Dick Smeeton (Admiral Sir Richard Smeeton, Director of the S.B.A.C., who led all the missions I have mentioned) that Lombard would deal with the more technical (engine) aspects of the Show (we were allowed cameras) and I should give a general appreciation of what we saw. The Show was well described afterwards, particularly in the American *Aviation Week* journal. It was held at a new airport, Domodedovo, which had only recently been completed but was closed to the airlines for two months to permit rehearsals for the Air Show.

The Show was completed strictly to schedule at about noon and we repaired to the airport building for lunch, with the usual accompaniment of speeches and toasts. On Sunday we went round the Kremlin, the churches and the armoury and also the Red Square with Lenin's Tomb and St Basil's Cathedral. That evening we went to the circus, which in Moscow is a nightly affair and, in the summer temperatures, has a rich scent of animals.

On Monday, the day of our departure, we went and bought Russian brandy, vodka and caviare at the special store for foreigners. We were due for an official meeting and then lunch with the Minister for the Aircraft Industry, Pyotr Dementyev, after first paying a call on the British Ambassador, Sir Geoffrey Harrison. Our meeting with the Minister was, as on previous occasions, most pleasant but not all that informative. Dementyev had probably been in his job in aviation longer than any other Air Minister anywhere. He was in charge of an aircraft factory before the Second World War and became the Deputy Minister of Aircraft Production at the start of it. With the German Army near to Moscow, he was one of those responsible for shifting by rail all the machine tools, equipment, material and workers and other personnel from the Moscow area to the Urals and Siberia. No mean feat. In fact, terrific.

We were joined at lunch by two famous Soviet designer/constructors, Tupolev and Mikoyan. We sat opposite them. I had, by way of warning, previously told my mission colleagues that, from experience, we would be involved in many toasts across the table and that they shouldn't try to emulate our hosts, particularly Tupolev, whom I knew from previous meetings and who, in his late seventies, could drink vodka and Caucasian brandy without appearing to be unduly affected. That luncheon was no exception, but with Mikoyan motioning Tupolev to go easy!

I remember an earlier meeting, sitting with Tupolev in one of his machines at a Paris Aviation Salon and both of us discussing, in very bad French, the various aircraft and engine developments while drinking vodka and eating chocolates – a formidable and sickly mixture. The only other person in the cabin was a silent man in some sort of uniform, said to be an interpreter. Tupolev ignored him and then asked me if I wanted the man there. I shrugged and he sent him out of the machine. Tupolev was at one time kept under house arrest by Stalin although he continued his work and had meetings with his colleagues. He used to recount this with some amusement.

We returned home in the late afternoon in a B.E.A. Comet, loaded with drink, as we were each given a bottle of vodka by the Minister upon taking our leave. During the flight I sat with Adrian Lombard and got a bit worried because he kept asking me the same question over and over again. Having earlier suffered a burst blood vessel behind one eye he was told by the doctors to 'go easy', but that wasn't in his nature. On the Wednesday after our return he rang me once or twice to check that we were not repeating ourselves and saying the same thing in our reports. Then, in the afternoon, his secretary 'phoned to say that he had been taken ill and was in hospital. In the evening he died. I liked Adrian and his going was a great loss to us all, particularly to his Company. We did not always agree on future engines, but he was a natural born designer and a good engineer. What more can be said.

Since retiring in 1968 my work as a consultant has allowed me to continue usefully to use my experience and keep active what brain I have, which is important and even vital to one's health. Aviation is a vocation, particularly to those of us who have grown up with it; so that when the time comes to retire it is as if one has reached the edge of a precipice and the drop facing you is the emptiness of suddenly reaching the end of a long but interesting journey. I am not attracted to gardening, so that could not take the place of my work as a hobby. I like the country and enjoy being in it, but it doesn't offer me any active participation. I played golf regularly between the two world wars. It has always been bad golf, even bad enough to enjoy playing it alone on a Scottish holiday course and to be with one's thoughts. More correctly, I should say that I play at golf. Walking never appealed and golf is the most pleasant and interesting way of taking a mild exercise – 'going for a nice walk' was an anathema to me as a child.

*   *   *

We and the Germans gave the Russians some aero gas turbine technology, ourselves freely and the Germans involuntarily. With the end of the war and the General Election of 1945 resulting in a Labour landslide, Stafford Cripps left the Ministry of Supply to become President of the Board of Trade. Our new Minister, John Wilmot, was approached by a Russian delegation who wished to order Derwent and Nene gas turbines from Rolls-Royce. The file came to me from the Minister's office asking for my views, which I gave to the effect that if we let the Russians have these engines we would be selling our birthright and they, buying time, would be saving themselves about five years of hard development. I was against the deal. But Cripps stepped in and Rolls were allowed to go ahead.

During the Korean War the MiG.15[1] had Nenes. The U.S. Navy salvaged one from the sea, that had been shot down off the Korean coast, and the engine was sent to the Philadelphia Navy Yard for examination there by the Yard and Pratt & Whitney. It proved to be a Russian built Nene rated at over 6,000 lb of thrust; in fact, about the same rating as the (larger) Tay.

It has been recorded by Yakovlev[2] that when the purchase of Rolls engines was suggested to Stalin he called it naive and said, 'What kind of fools would sell their secrets'! For a man of such brilliance, and the leading patents lawyer before the War, Cripps could be politically blinkered. He thought, being a Socialist, that when he went to Moscow as our Ambassador during the War he could influence the Russians and get close to the hierarchy. But they took little real notice of him and he apparently did not get near to Stalin. Again, in India, he was more Gandhi than Gandhi.

Credit should, however, be given to Cripps for his realistic appreciation of the Russo/German situation. He appeared to know that Stalin didn't trust Hitler, didn't trust anyone in fact, but would not seek to upset the Germans. He (Stalin) would play along with them. Cripps, though, was fairly sure that Hitler intended eventually to attack Russia. It has always been credited to us that Churchill warned Stalin that the attack was imminent, which he did. But it seems that Stalin was fairly well aware of this already, although when Richard Sorge, his chief spy (later executed by the Japanese), told him that Hitler would attack Russia with 170 divisions – 3 million men, plus – on 20 June, 1941 the warning was ignored. In the event Sorge was only two days out.

As a Minister, Cripps was good to work for and, outside politics, considerate. His lawyer's training showed to advantage when we, the Directors General, briefed him before a Cabinet meeting. He took our briefs home to read, but when he left for the Cabinet meeting next day he was word perfect. Further, when any of the senior staff in the Ministry got promotion or an honour, almost as soon as the man himself knew of it there was a handwritten note of congratulation, always in red ink, from the Minister, delivered by messenger. I myself received one when I was promoted to Air Commodore.

As for the involuntary giving of gas turbine technology by the Germans to the Russians, they of course had no option since the Junkers plant at Dessau and also a B.M.W. plant were in the Russian zone. One night there were knocks on the doors of the principal Junkers technicians. They were taken to a train, without time to get any personal or household things together, and transported to Kuibyshev, some 900 kms E.S.E. of Moscow. There they were met and taken to houses allocated to each technician and, eventually, his family. They worked on gas turbine designs for the Russians for about six or seven years, after which they were allowed to return to Germany. I met some as they came through West Berlin and the interesting thing was, they said, that they couldn't get even nuts and bolts made properly. Their experimental engines were long delayed in being built and getting to the test bed. It was quite apparent that the Russians were taking what they thought worthwhile from the Germans and handing the information over to their own engine manufacturers. Those returning engineers I questioned in West Berlin said that the Russians were unlikely to make a workable gas turbine. But since they were not told or shown anything, and didn't know of the Rolls-Royce deal, their state of ignorance was not surprising.

Today, from the little one really knows of Russian gas turbine technology, it is plain that while they may have some reasonably good military engines they have a long way to go to equal Western gas turbine technology, particularly for commercial aviation. Their more recent approaches to U.S. and British aircraft and engine manufacturers suggest the wish to have information on the certification procedures for modern commercial airline aircraft. Today, the small Yak 40 has been the only commercial aircraft to sell outside Russia.

---

[1] MiG stands for the names of its joint designers Anton Mikoyan and Mikhail Gurevich, both of whom trained under one of the early Russian designers, Polikarpov.

[2] *The Aim of a Lifetime* by Alexander Yakovlev. (Page 327).

21 *Some Encounters over the Years*

The man who interested me very much between the two World
Wars was Erhard Milch, whom I first met in the 'twenties in the
early days of Lufthansa and, through him, Von Gablenz and Drs
Stüssel and Schatzki. These men created and built up this first class
and world famous airline.

Even at that early period Milch was interested in fuel and it
could have been that his farsightedness set him thinking about
what could be done to ensure Germany's future fuel supplies in
case of a national emergency.

He was not popular because he could not suffer fools gladly. He
was too good an organizer and administrator, and too energetic,
to be particularly liked by some of the incompetents there were
around him in Hitler's hierarchy. There were, of course, the usual
Service jealousies directed at one who was not a career officer and
who had achieved such high rank. But Milch was respected by those
who knew their job and appreciated his competence. He could even
be sympathetic and considerate to individuals who had failed, either
through incompetence or ill health, and saw that they had material
help. But he was quite ruthless, and apparently became more so in
the War, where anyone's actions threatened the State.[1]

I can, of course, only speak of Milch prior to the War, but my
further meetings with him after he became Göering's Deputy and
during the build up of the aircraft industry and the Luftwaffe,
while less frequent, were always to the point and mainly dealt with
engine development progress. After 1936 I was mostly accompanied
during these meetings by Udet. Twice I was included in Milch's
dinner parties at Horcher's restaurant in Berlin, once when I was
dining there with a colleague when we were invited over to his table
and once as the result of a formal invitation.

A study of Milch is interesting. Why, for instance, did such an
intelligent and able man fall for Hitler? For the answer one must go
back to the Germany of the 'twenties and the aftermath of the

1914–18 War, when people were starving and there were revolutionary groups and parties creating disorder, with shooting and rioting in the towns. I was there quite frequently at the time, first in the North German ports in ships and later in aviation. Milch himself was at one time involved in restoring order. This situation coupled with the collapse of the Mark, causing the ruin of thousands of individuals and many companies, with millions of unemployed and the weak Weimar Government, cried aloud for a leader. Small wonder then, that all who wanted to see order restored and Germany rise again rallied to the only obvious leader – Hitler. But as Milch, Speer and others rose to high estate, they came to worship the Führer, even when he was going down the other side of the hill after the failure of the campaign in Russia and the Stalingrad débacle. Of course, the higher one got in the Nazi hierarchy, and the longer one was with Hitler, the more difficult it was to get out – except by being sacked, disgraced or executed! But Milch and Speer, both very efficient, remained loyal.

This inevitably brings me to Udet, whom I had previously known and liked as a person. He had been a fighter pilot in the First World War and afterwards became a master of aerobatics. I wondered how he would or could handle the big job he was given in 1936. His position at the start of the war was very similar to that of Tedder, who was Deputy Director of Research and Development under Sir Wilfrid Freeman, the able Air Member for Development and Production at the Air Ministry.

I worked with the German Air Ministry and the engine manufacturers, on and off, from 1933 to 1939, and naturally saw a good deal of Udet both from the working and social aspects. I can confirm, even at that time, the personalities involved both in the airframe and engine industries, and the Government departments, with their ambitions and jealousies. In too many cases incompetence made everything immensely more difficult and the situation literally needed the strength and ability of a Milch to control.

There was an incident in 1936, when my visits to Germany were fairly frequent: my usual practice was to work out with the particular aero engine company the pattern of testing for power and duration in order to prove the reliability of the exhaust valves. The company would run the test and then ask me to come and view the stripped engine. On this occasion, the company concerned was Siemens (later Bramo-Hafnir) in Berlin. I had been to them a number of times to advise on valve treatment and always saw the company head, Wolf, and his engineers. On this visit, the engineer who usually took me to see the engine wasn't there and another of his colleagues deputised. When I asked where he was (I forget his name), I was told that he was in Dachau. He was a Jew and so was Wolf, the head of the firm. The German Air Ministry had kept the S.S. away from them, since both Siemens and Hirth had been building their small engines for the elementary training aircraft which were used by the Luftwaffe at that time.

Upon hearing about Dachau, I went to see my friend Udet, who told me that they were losing out against the S.S. Goebbels was campaigning in his press against the Jews, anywhere and everywhere. Would I come and see Milch? 'Yes', I said, since I was quite angry. We saw Milch and I told him that, while not entering into German politics, I was not going to be very happy if this sort of thing was likely to happen to the people I was working with. Milch said, 'Mr Banks, you will be with us at the Haus der Flieger reception tonight. Hermann (Goering) is coming. Would you tell him what you have told us?' I said, 'Certainly'. We saw Goering who, in irritation, said, 'Ach, Himmler's lackeys again'. I heard on my next visit a month later that this man had been released and was in Switzerland, so I was quite gratified that I had taken the trouble to 'interfere'.

It is amazing to note the muddled state of the German aircraft programme more or less throughout the War and how long it took, say, to find out if a heavy bomber and long range reconnaissance aircraft (He.177) would even approach what the designers and manufacturers promised. More amazing is that the Messerschmitt company was in financial difficulties in 1942, with men laid off because of the mess made of the redesigned Me.210 and its necessary cancellation by Milch. It was revived, Phoenix-like, from Voigt's original design and became the successful Me.410. Then came the Me.262. But all so late, thank Heaven!

Hereunder is an extract from Wheeler-Bennett's book which is I think interesting:

'Adolf Hitler had an uncanny understanding of the German national psychology. He recognized, whether by instinct or study or merely by a fellow-feeling, that the German people are the most inhibited in Europe, fundamentally governed by a sense of inferiority for which they over-compensate by arrogance, and altogether lacking in self-assurance and a sense of responsibility to themselves. All these symptoms Hitler read aright and used them to his own advantage.'

*   *   *

During the summer of 1936 I had an interesting visit. I had been seeing the power plant section at Wright Field and left in the afternoon for a small meeting in Dayton, attended by Dr George Lewis, the able Director of the N.A.C.A. Aircraft Engine Research Laboratories (later, the Lewis Flight Propulsion Laboratories) at Cleveland, and Lester Gardner, Secretary of the I.A.S. (the Institute of the Aeronautical Sciences, now the A.I.A.A.). They wished to have a talk with me on European matters and were later to go and see Orville Wright. In fact, they had already asked him if they could bring along an Englishman. It developed into a most interesting talk with a dignified and quiet man, who was still very interested in all aspects of aviation. When he learned that I knew Europe and, particularly, visited Germany frequently, he wanted to know where they were going and what was their aeronautical capability. He appeared most interested to hear that my father had got his brother, Wilbur, his first mechanic when he went to France.

*   *   *

Looking back, some of my more lively trips between wars were in and around Europe, some of them to Yugoslavia, Zagreb and Belgrade visiting oil refineries, Ethyl blending plants and the Air Force authorities. I often accompanied Michael Golovine and H. K. Jones, who were representing Rolls-Royce and Hawker respectively. Michael was then the Rolls installation engineer, having changed over from selling Rolls cars to high society in Paris. Hawker were in the process of interesting the Yugoslavs in the Fury fighter, which they eventually succeeded in selling to them, also negotiating a manufacturing licence. Some of the visits I made were in order to 'dope' the fuel so that 'George' Bulman, the Hawker chief test pilot, could demonstrate the machine at full throttle. Later, I went there to 'dope' the fuel for the Yugoslav Fury entry for the 'Round the Alps Race' which it won, piloted by Captain Shintitch at an average speed of 201 mph. A good friend, Tom Mapplebeck, used to entertain us in Belgrade.

I had a bad experience on one visit to Zagreb. I had gone there after visiting some oil company head offices in Genoa and was to inspect one of the refinery blending plants outside Zagreb for Standard Vacuum. I began to feel very ill before I left Genoa, with a headache, a temperature of 104°F and sweating and aching all over. I managed to get to Zagreb on the Saturday and continued to feel so bad that I went straight to bed. On Sunday, a doctor was called in and he diagnosed poliomyelitis! My temperature went to 105° so he gave me something to increase the sweating and also a tablet that tasted very bitter, like quinine. In about 48 hours the temperature went down and the sweating eased. I was walking about again in a few days, but very weak. Whether or not it was polio I don't know, but I have rarely felt so ill. If it was polio then I was very lucky not to have become incapacitated in some way.

In Zagreb I could get by with my German but in Belgrade the only other language was French, which was more difficult for me. But our Yugoslav friends with whom we dealt mostly spoke fair English. I visited Novisad, the Air Force and technical headquarters and also the Zemun airfield across from Belgrade, to deal with fuel and engine problems.

We had many good parties in Novisad and Belgrade and were well entertained by the Air Force and technical personnel, drinking considerable quantities of their potent national drink, Slivovitz, a plum brandy. One night in Belgrade, Michael Golovine and I found ourselves helping to entertain a troup of Greek dancers. It got a bit wild and out of hand when some Yugoslav Air Force officers were instructed by the girls how to put on a dance show.

The journey from Zagreb to Belgrade was by train, and in those pre-war days there was a Customs inspection upon arrival at Belgrade, not so much concerned with dutiable goods, but mainly to search for arms. On every occasion, then, when an important foreign diplomat or a leader from another country visited the capital, the police stopped all public transport, 'buses and trams, during the arrival and departure periods, because of the fear of bombs and assassination attempts. All the Belgrade hotel hall porters were either police spies or informers.

*   *   *

Another crisis occurred some years later, in 1958, when I was in Cleveland, Ohio, visiting Thompson Products (now T.R.W.) and the White Truck Company. This was on the eve of my sixtieth birthday, Friday, the 21st March, and I was due to have my birthday dinner with Jack Macauley in Washington the next day.

I had been dining with Bob Cass of White, and Thompson's had sent me to his house in a large limousine which came to take me back to my hotel after dinner. It was about midnight when I left Harriet and Bob, and I dozed off to sleep in the back of the car. I woke up in hospital on what looked and felt like a mortuary slab, with two young interns staring at me. The time was about 1.30 a.m. and I had been unconscious for over an hour. Apparently, the driver was making a left turn at an intersection, following too slowly behind another car, when the lights changed and we were hit in the middle of the turn by a car coming in the other direction without slowing up. It struck us at the near (right hand) side back seat where I was sitting, projecting me violently across to the other side of the car. The interns pressed my chest and then strapped me up around the chest and ribs. My right side looked somewhat depressed. They said I was only bruised and, anyway, as there was no x-ray available at that time of night they suggested I go back to my hotel. The hospital I was taken to was, according to police rules, the nearest to the accident; but among all the good hospitals in Cleveland this was probably the worst, notorious mainly as a home for geriatrics and alcoholics.

When I arrived at my hotel, my side hurt so much that I could not go to bed and had to sit in a chair. I slowly packed my bag, which took most of the night, and boarded the limousine for the airport next morning. Arriving at my hotel in Washington, I rang Jack and later went and had my birthday dinner with him. Returning to my room at midnight, I still could not lie down and had to prop myself up, sitting, with pillows. On Sunday, I called in the hotel doctor who thought I was in a very serious condition and should come for an x-ray examination on Monday. At the first count, he found about eleven fractured ribs with some compound breaks. The final count was thirteen rib fractures and a crushed lung, with haemorrhage and debris (fluid) in the chest cavity. The doctor advised I go straight to hospital, but I said I would prefer to get to New York where I had a number of friends. I didn't tell him that I was due for dinner at the Links Club on the Tuesday evening, to say farewell to a good and respected friend, L. S. (Luke) Hobbs, Vice-President of engineering of Pratt & Whitney, upon his retirement. It was a large but intimate dinner, attended by many aircraft people and other close professional friends of Luke's. Included were Fred Crawford and Arch. Colwell, President and Vice-President of Thompson's, who had been to Australia and had flown direct to New York from Los Angeles. They had only heard by telephone of my accident in their company car. When, however, they saw me at the dinner, Arch. immediately left the room, spoke to a surgeon friend, then pulled me out of the room and rushed me to the Roman Catholic St Claire hospital, where two of New York's leading medical men, a chest surgeon and a chest physician, were to look after me for five weeks. It took them about two weeks to get my chest thoroughly drained of debris, and give me a number of blood transfusions. But I had a relapse at Easter weekend. Victor De Luccia, the surgeon, had said to me on Saturday that he hoped I wouldn't want him as he wished to have Easter Sunday with his family at Yonkers. That night, while reading at about midnight, I felt a very heavy pressure in my chest and was getting short of breath. I got out of bed and sat in a chair, but felt worse. I just managed to press the emergency bell before passing out. When I regained consciousness, I found myself surrounded by interns and nurses, Lebanese, Greeks Negroes, Irish and Mexicans. Apparently, the edge of a yet unhealed rib had cut into my lung causing severe haemorrhage into the chest, and the flooding pressed on the heart and stopped things. When the heart stopped the blood flow stopped therefore the pressure on the outside of the heart reduced and, luckily for me, my heart restarted. Poor De Luccia and Sydney Bassin, the physician, had to come in at 2 a.m. on Easter Sunday morning and drain me, using the long, six inch, hollow needles to penetrate my chest. Then they said they would come back in the middle of the morning and give me blood transfusions, since they had taken out over three litres of blood and debris from me. But I said, 'No'. Please would they fix the transfusions right away and then go home, leaving the hospital staff to complete them, as there were some five litres of blood to be given. After all that, I had three nurses on eight hour shifts to look after me.

Without doubt I was saved by these two capable medical men; after five weeks in hospital they allowed me to go home by ship, not by air. Con, my wife, had been flown to New York to be with me after this crisis. She couldn't come before since our younger daughter was in hospital in London and couldn't be left. We sailed home in the *Queen Mary* although I would rather have flown home since I still could not lie comfortably and the air journey would have been shorter. To expand the collapsed lung Sydney Bassin made me do breathing exercises twice a day for twenty minutes a time. It was this treatment that got my lung right and, today, I only lack about 500 ccs vital capacity out of, I suppose, some 2,000 ccs. As the years mount elderly people don't breathe so strongly; they more or less hibernate!

'Kelly' Johnson of Lockheed, whose supersonic and near hypersonic aircraft (F104G and SR71) engineered by his team in Skunk Works are so famous, amused me when he had returned to work after a serious abdominal operation. Talking about it he said, 'You know, Rod, they cut some weight off me too, and it cost about the same as the weight reduction we did on the C5A (Galaxy) – $500 per lb'.

---

[1] *The Rise and Fall of the Luftwaffe* by David Irving, published by Wiedenfeld and Nicholson.
[2] See *The Nemesis of Power* by John Wheeler-Bennett.

22 *A look at Engines, Engineering and the Future*

Six decades of working with marine diesels, aero and automobile engines have given me some insight into the attitudes and outlooks of various firms.

There are still examples, but now much fewer, where Design does not invite Production in at the start of a new project, to ensure and endorse the best possible detail layout for economical manufacture. Between World Wars, Roy Fedden (the late Sir Roy Fedden), as chief engineer of the Bristol Engine Company, rightly, used to praise Whitehead (head of his special, pre-production, manufacturing section) who could make anything the design office gave him. My comment to Roy was that if this undoubted genius, and he was, would throw some of the designs back for rehashing and simplification he would have rendered a still greater service to the company. I should add that when 'Bunny' Butler was alive he and Whitehead were complementary in checking the design of engine components for producibility. But without Butler there seemed less attention in this respect, though admittedly engines *were* becoming more complex.

An important relationship is that between the parent firm and its component and equipment suppliers. This should be considered on a long-term basis where continuing business between them results in improved components and equipment, say for a cooled turbine blade or a complex fuel control system. Improvement in this sense means quality maintenance and, with more background, useful innovation. In other words, the parent/supplier association should not only be based upon inspectional acceptance of the suppliers' products but on mutual trust developing into continuous and continuing relationship.

I have, incidentally and on occasion, noted the arrogance of a parent company in its treatment of a reputable supplier, rather like a tourist haggling over a piece of brassware or a carpet in an Arab Souk.

Then comes the question of quality control, generally evolved as a small department concerned with the maintenance of component and unit quality having a head or controller responsible to a senior engineering member such as the technical director. Alternatively he may report to a chief engineer brought up in the old fashioned chief engineer's way of working through all departments in his professional training and subsequent positions.

The existence of a quality control department headed by an enthusiastic controller has often tended to lead to an autocratic and dogmatic attitude and separation from the rest of the company organisation – too much 'interference control', in fact. But there were times during World War II when we had to have an autocratic quality controller, for example when Rolls-Royce and my department in the Ministry decided that quality control of radiator and oil cooler production was essential. Rolls then provided one, Frank Nixon, to act for both of us. He did a magnificent job and really brought quality into this important equipment.

Quality control should not be confused with inspection. When official aviation started in this country at the Royal Aircraft Factory (now R.A.E.) under Lieut. Colonel M. J. P. O'Gorman, its head, there was created on the 2nd January 1914 the Inspection Department of the Military Wing, R.F.C., under Major J. D. B. Fulton. This was to develop into the A.I.D. (Aeronautical Inspection Directorate) which was for many years a kind of combined inspection and quality control organisation, headed first by Lieut. Colonel Bagnell-Wild and then Colonel W. Outram. Its inspectors were posted to the firms in the aircraft industry and had materials testing headquarters and laboratories elsewhere. The A.I.D. was particularly good on engine matters and dealt mainly with aircraft constructional stressing but not aerodynamics. As things progressed the individual or parent engine and aircraft firms were 'approved' to act as their own official inspectors, provided they had a full inspectional organisation meeting A.I.D. requirements. In recent years, the A.I.D. has been replaced by A.Q.D., meaning Aircraft Quality Assurance Directorate; almost the same thing without, perhaps, the almost autonomous role of the original A.I.D. in its heyday.

I have mentioned how the U.K. parent company, Rolls-Royce, did more 'in house' manufacturing than was the practice in the U.S. This is perhaps because of lower scale ordering and, therefore, shorter production runs than in the States. However, it cannot be over-emphasized that, in spite of shorter runs, it pays the parent company to keep a healthy sub-contracting organisation going without, of course, too greatly denuding itself. A pocket or cadre of skilled labour is then created available to cope with short notice demands for increased output or a change in the engine specification, a modification which might normally make heavy demands upon the parent.

The American engine manufacturers find the established component firms, like T.R.W., of considerable help and they have become part of the parent firms' working capacity. This system also finds favour with the U.S. Armed Services, to meet a national emergency and to preserve and spread the particular engineering skills.

Aviation, particularly commercial aviation, is a risk business and being a risk business it is all the more necessary to have experienced professionals, professional engineers and engineering and airline analysts, to determine policy i.e. the right engine to build which, to ensure adequate sales and production numbers, must also be suitable for at least two and preferably more different makes of airframe.

On the subject of diversification: I have always held that one cannot truly diversify in the aircraft industry if one is intending to keep very active in aviation. The alternative is to get out altogether. However one of the more successful diversification attempts or change-arounds in recent years has probably been that of the Hawker Siddeley Group, where the changeover (reduction) from aircraft to industrial engineering has been from, roughly, 60 per cent aviation and 40 per cent industrial to the reverse. The Group diversified into moderately heavy industrial engineering i.e., large and small diesels and electrical equipment. The Garrett Corporation (Garrett-AiResearch) of America has also been extraordinarily successful in advancing into other fields of high technology by applying their aerospace experience and techniques.

Since the aircraft and its engine takes the maximum of skilled manpower for one pound weight of finished product i.e. the maximum in labour cost, compared with the bulk of consumer goods; even motorcars could not in any way replace aircraft in an aircraft manufacturing environment and organisation. So one is either a success in aviation or a failure; though there has been, to my mind, too much drifting along by some firms which, but for the misguided idea of insisting upon competitive contracts imposed by the Government Department concerned, ought to have gone under long ago.

The large commercial aviation engine could be a doubtful proposition to all but the strongest and most successful manufacturers since the wide-body, fanjet, passenger aircraft is going to last for possibly up to twenty years in first line service. Which means that, while additional aircraft will be ordered from time to time, the four large engine manufacturers, General Electric, Pratt & Whitney, Rolls-Royce and S.N.E.C.M.A., may not all be overburdened with orders at any one time. In the military engine field, reasonable numbers may be ordered. But in the U.K. and also in the U.S. there are new aircraft presently under development which, when they reach the Services, will continue into the late 'eighties or even the next century. There will be modification phases to meet different rôles, but not many requirements for entirely new aircraft. In the world political scene, however, America may always be somewhat committed to consider new requirements while keeping level with the U.S.S.R.

I have, with others, been irritated by the writings of the uninformed on the Concorde. This aircraft is an experiment but has now become a fact of airline life. Its debut could perhaps with benefit have been hastened by getting it into airline service a year or two earlier. But that is my only criticism. Those who have written and spoken at length against it had no facts to go upon because, being an experiment, they obviously could not know, nor were qualified to understand, the problems involved in the design and development of such a machine.

I have often wondered about the true economies of the wide-body commercial fanjet. Since aircraft are designed to overcome gravity minimum drag allied to reasonably high power/weight ratios are important. For long ranges this could point to the supersonic aircraft rather than big volume, subsonic machines; large they must be to cater for the passenger load, with a lot of wasted internal airspace as a bi-product. The slim supersonic machine having a small volume cabin, with comfortable seats but the minimum of cabin luxuries (elaborate food), could, in the end, with high frequency and speedy turn-arounds, prove an economical way of transporting passengers for long distances. The wide-body, subsonic, high passenger capacity fanjet aircraft tend to swamp airport facilities, but nevertheless they are suitable for relatively short inter-city and domestic (subsonic) flights. It is to be hoped, however, that passenger capacity limit will not exceed, say, 500 – from insurance and safety viewpoints.

L'Affaire Rolls-Royce

Probably the greatest national shock in recent years was Rolls-Royce going into voluntary liquidation because they had not the ready money to pay the salaries and wages. It has always seemed curious to me that a company, forty per cent or thereabouts of whose work was, and indeed is, on Government contract, should not be able to call up some progress payments or be allowed a non-interest Treasury loan to meet the immediate crisis and carry on the development of the RB211 engine for the Lockheed TriStar. When all is said and done, this had become a national commitment but it suffered poor handling by the Government when spelling out their 'lame ducks' policy. Sheer panic did the rest; but it was unthinkable that we could 'rat' on the R-R/Lockheed contract and hope to hold our heads up again in America.

A week before the decision was made to put Rolls into liquidation, a full meeting of the new Board (not the present Board) was called to discuss the situation. This was attended by Air Chief Marshal Sir Peter Fletcher (Controller of Defence Procurement at the Ministry of Defence) accompanied by a senior director of the Directorate of Engine Research and Development. Fletcher tried to explain the difficulties attending the development of the RB211 and, from the advices of Sir Stanley Hooker, how it could be put right. But the non-technical Board Heads seemed unable to grasp the significance of what he was saying – that the engine could be saved. At another Board meeting a week later, this time attended by the Minister, Lord Carrington, with Sir Peter, the decision was finally reached to put the Company into liquidation – the Prime Minister's edict, when the Conservatives came into power, being that 'lame ducks' must seek their own salvation or go under. Since those appointed in place of Pearson and Huddie had little or no experience of aero engines, how they were contracted for by the Government, and their long period of development, they acted from their purely commercial background. From those two Board meetings, no-one had apparently warned the Rolls-Royce representative in New York as to what was happening. And Dan Haughton of Lockheed, on his way to the U.K., was only to be informed of the liquidation when he arrived! As soon as the company went into liquidation, the Director of Contracts cancelled all the contracts on Rolls, which had to be re-negotiated – at a price!

Ignoring all the subsequent meetings and talks to decide the fate of the RB211, it can now be said that the two men who virtually saved the engine were Sir William Cook (Chief Scientist at the Ministry) and (Sir) Stanley Hooker who, in collaboration, decided that its continuance was worthwhile. Being the technical people they were, these two had the only knowledge among the hierarchy to give a considered and acceptable opinion.

Whether or not Rolls should have entered this toughest (airline) business, requiring an engine in the largest category, is the question. But the management considered their future had to be in the large airline business and in the largest market place, the U.S.A. That the cost of development of the RB211 was grossly under-estimated, or understated, can hardly be disputed but, here again, it could have been advanced that, 'we would not have got the order had we raised the price'. General Electric, the only one of the three major engine firms involved who had a prototype engine in being (for the Lockheed C5A Galaxy military transport), and who therefore had some basis upon which to estimate a likely cost and price, pulled out of the competition because they could not compete with 'that kind of Dutch auction'.

To sum up: Rolls-Royce's liquidation in 1971 was not basically a failure in engineering, though they did get somewhat oversold at the start of the RB211 on the 'Hyfil' fan (their carbon fibre fan-blade material) and went on too long with its development. While the two top men were engineers by training and experience they, like all of us, could make mistakes – not so much of ignorance but rather of an over-confident approach. It was the happenings subsequent to the crash that, to me, were rather pathetic. Beyond the appointment of (Sir) Stanley Hooker to the Board just before the liquidation, there was no full-blooded attempt to re-establish the firm's engineering management as a top policy-making group or organisation. This continuing lack of a strong engineering group, as distinct from capable technical individuals, is still worrying to those viewing the future prospects for our top and only large engine builder. In fact, today, there are only two builders of large engines in the Western World run by engineers – General Electric and S.N.E.C.M.A. Energetic chairmen, however experienced in other quite different disciplines or businesses, are only able to give moral encouragement to the engineering members of their Board and to monitor what is happening through them and the financial director. He, the chairman, cannot effectively discuss with his engineering colleagues and their staffs technical matters upon which to base future prospects, though admittedly he may eventually acquire a feel for the right decision.

I felt strongly against the takeover of Bristol-Siddeley Engines by Rolls-Royce because I did not think, in becoming such a large aero engine manufacturing unit, that Rolls would generate any more business to warrant the increase to a size even larger than the most successful manufacturer in the World, Pratt & Whitney. Pratt & Whitney, like Topsy, grew to its present size by working up to it in the commercial and military engine fields before the Second World War, and during the war by an enormous increase in military engine production. In the aftermath of war, they had large domestic and international commercial demands to satisfy, with an appreciable upward military production kick when the Korean War came followed by the Vietnam affair. We in the U.K. had nothing like this build-up after World War Two, nor have we the size of domestic market upon which so much depends in order to launch a project successfully.

The RB211 could have been offered by Rolls without the Bristol tie-up. Their fear was, apparently, that Bristol would do a deal with Pratt & Whitney for the JT9D. This was well founded, but Rolls at least won the TriStar contract as much on technical acceptability and innovation as on (too) low price.

With Rolls-Royce and Bristol Siddeley as separate and separated engine manufacturers, there at least remained an element of competition even, perhaps, offering more scope to deal with demands for a greater variety of engine types.

The airframe industry, even without nationalization, has too often experienced dithering and lack of official conviction when dealing with a Government requirement (contract) and projects have been delayed and killed after much money has been spent, e.g., the TSR2. This aircraft had, however, suffered early changes in requirements, resulting in a confusion of rôles and cost overrun. There was also the case of the Hawker 1121, not officially supported but designed on a p.v. basis for eventual official support. It would have been a good supersonic follow-on to the Hunter, avoiding the later need for ordering the Phantom with a considerable saving in dollars and giving work to many here. But the damage was done by the Sandys White Paper of 1957, which seriously delayed consideration of aircraft operational requirements because of misguided ideas about rockets and missiles, etc.

To Rolls-Royce can be credited a number of technical advances on their piston engines in World War Two, which improved efficiency and performance. There was the ejector (backward facing) exhaust system which gives an appreciable increase in equivalent thrust horsepower at altitude; the supercharger, which was greatly improved in efficiency by Hooker after being earlier developed by Jimmy Ellor; also the ram air intake, giving increased boost due to the forward speed of the aircraft by converting kinetic to pressure energy. This was demonstrated in 1929 and 1931 on the Schneider 'R' engine. Then there was the flame trap, to prevent blow-back and dangerous explosions at high intake (blower) pressures due to faulty sparking plugs, leaking inlet valves or weak mixture strength. The flame trap worked on the same principle as the Davy safety lamp for miners. Finally, another Rolls innovation was the liquid-cooled intercooler (or aftercooler), between the supercharger and the cylinder blocks, to lower the charge temperature by some 100°C.

Rolls designed and built a number of different engine types between wars and during the war. There was even a seven cylinder swashplate engine in 1923, probably intended for a motorcar, since it was of 3 litres cylinder capacity. Then came the X 16 cylinder in the 'twenties. There was a wooden model of an inverted 12 cylinder engine, named Merlin, in 1933, which was seen by a group of German engineers from Daimler Benz and Junkers when visiting Derby. But Rolls discarded this arrangement and the prototype Merlin, the P.V. (private venture) 12, was normal way up. Then there was a sleeve valve Kestrel built, but not proceeded with. In 1935 Rowledge designed the air cooled X, having 24 cylinders at 90° (with sleeve valves) of 4.2 in × 4 in giving 22.1 litres capacity and rated at 920 bhp at 11,000 feet. It first ran in September 1936. The Vulture for the Avro Manchester bomber came in 1939. This was an X, a double Kestrel of 42.48 litres cylinder capacity having a take-off rating of 1,800 bhp. There were various development troubles with this engine on the test bed and in the Manchester, and it was discarded in favour of the Merlin in the Lancaster. Finally, there was the Crecy sleeve valve, two-stroke, in development by Harry Wood during the War, but it necessarily had a relatively low priority. It presented a number of mechanical problems. The Crecy had a bore of 5.1 in and stroke of 6.5 in giving a cylinder capacity of 26 litres and a design rating of 1,400 bhp, but gave about 2,000 bhp when finally abandoned. The last new piston engine built by Rolls was the Eagle, a sleeve valve, 24 cylinder, horizontal H engine, à la Sabre, of 5.4 in × 5.125 in, giving a capacity of 46 litres. It had a two speed, two stage, supercharger and gave 3,500 bhp at 22 lbs sq in boost.

In recalling all this I am indebted to a Paper[1] by that incomparable development engineer, the late Cyril Lovesey. Incidentally in this Paper he quoted (Sir) Stanley Hooker who described the four-stroke engine as, 'one with one stroke to produce power and three strokes to wear it out'!

[1] *Milestones and Memories from 50 Years of Aero Engine Develpment* by A. C. Lovesey. Sir Henry Royce Memorial Lecture, Royal Aeronautical Society, 7th November, 1966.

23 *General Comments*

Life today can be likened to Monsieur Levassor's description of the first crash motorcar gearbox he designed for the Panhard et Levassor car in the early part of the century: 'Ça marche, mais c'est tres brutal'. Life also is tough and, in many respects, can be brutal. In some ways, I think the French and British are not unlike, particularly in regard to their attitude towards other countries. Though not islanders, the French can be as insular and, perhaps, just as intolerant or even more so. In military aviation, both countries are chauvinist in outlook and each considers it has the superior or obvious aircraft. But even so the French have, on occasions, been willing to look at promising foreign aircraft and improve on them. This is well illustrated by their very successful development of the Dassault Mirage, which had its start when the founder of that famous firm saw the record-breaking Fairey FD2 experimental, supersonic machine at the Centre d'Essais en Vol at Bretigny in 1956 or Cazaux near Bordeaux. Dassault was then reputed to have remarked to his able team that that was the proper shape to develop. So came the Mirage, which has generated considerable international sales. Why we, ourselves, didn't take the FD2 as the basis for a supersonic fighter, I don't know.

Referring to the French and particularly their high education, I am indebted to my friend and colleague, Michel Hedde, for the following: this stems from educational establishments like the École Polytechnique (in slang, 'X'), created by Napoleon for the training of artillery officers and having considerable influence in those ministries handling technical or technological matters (see Appendix I). Top civil servants so trained can more readily understand and deal with technological problems and discuss them intelligently with industry, whereas the British higher civil servants of the secretariat, who are university trained Treasury appointees, have little or no technical knowledge. I am not referring to our fully qualified scientific and technical grades who are, unfortunately, mostly secondary in position of authority to those of the Treasury appointed secretariat. They are really advisers.

The French Air Ministry, compared with our own procurement executive always gave me the impression of being more compact and decisive than the British Air Ministry between World Wars. There are now the so-called 'énarques', a new 'meritocratic' elite who are the graduates of E.N.A., theÉcole Nationale d'Administration. Since the Napoleonic reforms, the important positions in the French civil service have been held by members of the *grands corps* who have passed stiff competitive examinations prepared by the special university-level grandes écoles – the non-technical level civil servants going to E.N.A. It might appear that the French system is democratic and egalitarian, but until 1945 it was not. Before then fees were paid to go to these écoles, but in 1945 they were nationalized. However, even this has not allowed anyone and everyone to be accepted into, for instance, the 'Inspection des Finances'. About 60 per cent of all the top French civil servants come from the prosperous middle class and nearly 30 per cent of the 440 top civil servants between 1946 and 1969 were drawn from only three famous *lycées*, all in the upper middle-class districts of Paris.

That the British have education and the French learning remains true in the broader sense although the British public school product, if not worked so hard, also sees something of the social side of life. But, today, French learning appears better able to meet the higher needs of the industry, the civil service and the armed services for up-to-date and advanced knowledge. In speaking with my French friends, I have found, however, that they have considerable respect for the British civil servant, insofar as his basic education and deportment are concerned.

*   *   *

I have always disliked large meetings, overlarge committees, etc., and well remember two such in the last war. The first was held in the Ministry of Supply after Wavell's successful desert campaign when he defeated and captured the Italian army. In the battle, Wavell lost a number of armoured vehicles, tanks and half-tracks, which was normal, but a number just 'stopped in their tracks' for no apparent reason. The M.o.S. meeting was chaired by a General and attended by the armoured vehicle technicians, including the two senior engineers of A.E.C. and Leyland who had manufactured the engines, Rackham and Spurrier. Also present were two officers who had been armoured vehicle commanders in the desert. I was asked over from M.A.P. to listen and contribute if needed. The Chairman rambled through the reasons for the meeting, with many diversions, for more than half an hour and never once referred to the two officers who had fought with the vehicles and experienced the stoppages, until Rackham, thoroughly fed up, murmured to me that he was going to ask the Chairman if he could question the officers. He then asked them why they thought the engines had stopped and the senior, a Captain, said 'Oh that's simple, Sir. You see, the engines are flexibly mounted and the fuel pipes are rigid and fracture'. The meeting then broke up!

The second occasion occurred about a year later, after my assistant director (technical) on propeller development had come into my office while I was having a small meeting and whispered, 'Could you take a meeting on Rotol propeller blades in about ten days time'. We had been having trouble, particularly in the hot climates, with the high density wooden blades shrinking and loosening in their steel hub adapters.

When the day of the meeting arrived, my assistant director ushered me into a room filled with some forty or fifty people. They were mostly engineer officers from all the commands in the East, Middle East and Home. I said quietly to my very good A/D that if he did this to me again, I would sack him. I took the chair and the meeting lasted five or six minutes, after I had told the assembled company to choose a panel of six from their number and report the facts and their recommendations in twenty-four hours.

*   *   *

Coming back to engines: I gained a great deal of knowledge and experience of the piston engine, having spent from about 1912 to the end of the Second World War intimately concerned with its 'hot' or combustion end. As a result I became particularly knowledgeable on exhaust valve design and the high-temperature, corrosion resistant, austenitic steels used for its construction, along with valve seat and head facing materials such as 'Stellite' and 'Brightray'. Among the less glamorous but nevertheless important problems to come my way were sparking plug suitability, design and behaviour; and the detail design of liquid and air-cooled cylinder heads.

Developing from my marine diesel background, I was also interested in engine lubrication and bearing design, particularly after experience with Michell's clever application of Osborne Reynolds' original wedge theory relating to lubrication and lubricating oil films. The Michell thrust bearing gave to marine engines a compact, low weight and operationally efficient means of absorbing propeller thrust, which eliminated the old multi-collar or horseshoe type of thrust bearing that had previously been universally used. At that time, between World Wars, I was often surprised that engine designers, and particularly development engineers, did not always appear to appreciate this theory of bearing lubrication – at least judging by the oilways that used to be cut all over a bearing face, so preventing an unbroken pressure film from building up.

The Old and the New. Top picture *Old type 'horseshoe collar' marine engine thrust bearing*. (By permission of the Institution of Mechanical Engineers). Lower picture *Michell marine engine thrust bearing of modern design*. (Courtesy of Vickers Engineering Group).

A good illustration of this was the case of the floating bush at one time fitted to the master-rod of the Bristol Pegasus engine already mentioned. This bush, drilled with many holes to pass the oil from the inner to the outer surface, needed a relatively large oil circulation which did not, however, create a high pressure film build-up. When Senkowski of P.Z.L. managed to prevail on Roy Fedden to return to the fixed bush a great improvement resulted.

Heron's brilliant work on cooled valves, when the sodium filled exhaust valve became universally used, permitted more power per cylinder and gave longer valve life. The earlier days of the sodium cooled valve brought problems of excessive valve guide wear and even red hot valve stem tips. It also took the temper out of the springs, all due to the increased heat from the valve head overloading the guide. Cylinder head cooling immediate to the valve was often inadequate and did not allow the heat to dissipate effectively. Hence the very sophisticated close-forged American cylinder head with 'gashed' close pitched fins backed up by the N.A.C.A. cowling.

All these problems, and many others, were only surmounted by intensive development, the real essence of creative aero engineering.

I like professionalism in any and every pursuit and, indeed, the world now demands it from the individual and groups and organisa-

Author with John Mayo and the Bristol 407 car he keeps like a new pin.

tions. For instance, John Mayo, who has driven me for many years, is a true professional and knows his car intimately. He was Rolls-Royce trained and can effect maintenance and running repairs. My car, the first Bristol 407, has now done more than 300,000 miles and looks completely new, including the engine which is always kept polished.

24 *Courage*

Four of the most courageous men I know, or knew, since two are dead, I number among my French friends: Jean Allais, Claude Bonnier, Gerard (Gerry) Morel and William Savy. I can do no better than quote from the reference below which relates to 'Gerry' Morel,[1] who died in August, 1960, directly as a result of his wartime service. Quoting from the book:[1]

'Probably the bravest man to serve S.O.E. (Special Operations Executive): Major 'Gerry' Morel was Liaison Officer to a British Regiment in the Spring of 1940. He was captured on the beach at Dunkirk, having missed the last transport when lying seriously ill and exhausted on the ground. The Germans released him later from prison camp because they thought he was too ill to survive for very long. Hardly recovered, Gerry went to Spain and, knowing nothing of S.O.E. but determined to fight again, he signed on a ship going to Brazil, hoping to reach England somehow. From Rio de Janeiro he returned to Portugal, contacted British Intelligence in Lisbon and finally landed at Baker Street. On the 4th September 1941, he was one of the first S.O.E. agents back in France. His poor health precluded parachuting and he was the first man to be landed by Lysander. In fact, since the Fall of France, this was the first time that an R.A.F. 'plane had landed in France. He had landed to learn what had become of Georges Begue, an early radio operator. While trying to establish radio contact with headquarters, Gerry was betrayed after six weeks, taken by the Vichy police to the prison at Perigueux and went on hunger strike and again fell seriously ill. Removed to hospital at Limoges, he was operated on and, with the stitches still in his stomach, escaped from prison hospital.'

He got to Lyon, where he lay hidden with his friend, Jean Allais, who had been with Citroen in Paris and had likewise escaped to Lyon. Both were in hiding and had to sleep together in the same bed for six weeks, Gerry with the still unhealed stomach wound.

They then got to the French-Spanish border and crossed the Pyrenees, on foot, in January 1944; those who have not seen this range in the depth of winter can hardly imagine what they went through. They were caught by the Spanish frontier guards and interned at the Miranda de Ebro camp, but managed to get away and reach Gibraltar, being flown from there to London.

William Savy used to specialize in getting people wanted by the Gestapo out of France to London, and he showed the same cold courage in being landed back in France by Lysander. He frequently dined at Maxims and listened to the conversations of high German officers.

Claude Bonnier and I had many contacts before the war. His work, like mine, was on fuels and he was based at the French national fuel testing laboratory which was, incidentally, located in Isadora Duncan's (the classical dancer) old house at St Cloud. In 1944, when the Germans were being pushed back, Claude turned up at my office in the Ministry. I was surprised and pleased to see him because we had lost contact after my last visit to France in 1939, when we had had a talk on fuel supplies and quality and afterwards dined together. He told me that he was in London for only a few hours and that it was urgent that he got back to France – Paris, I believe. Two weeks after he had left, word got back to me that Claude had been killed, shot by the Gestapo in trying to escape.

I was very pleased when, after the War, the French named a fuel test laboratory after Claude Bonnier.

With the French services in London was 'Colonel Vernon' a chief of staff to General Koenig, commanding the French Forces of the Interior. Colonel Vernon was, in fact, General Henri Ziegler, later to be head of Aerospatiale, builders of the Airbus and Britain's partner in the Concorde project.

[1] *Inside S.O.E.* by E. H. Cookridge, published by Arthur Barker Ltd., 1966.

25 *In Conclusion*

DINNER
following the
PRESIDENTIAL ADDRESS
given by
Air Cdre. F. R. Banks,
CB, OBE, HonFRAeS, RAF (Ret'd)

WEDNESDAY, 9th JULY 1969

ROYAL AERONAUTICAL SOCIETY

Chris Wren cartoon of the Author on the occasion of a dinner given by him on becoming President of the Royal Aeronautical Society after his retirement from Hawker Siddeley Aviation.

I have attempted in this autobiography to write without frills. In other words, when describing what happened, say thirty, fifty or sixty years ago, I have tried to avoid hindsight which can be very infectious and misleading in describing one's alleged reasoning after a large time interval.

In thinking over my life, I have come to the conclusion that the primary education of the young at the early part of the century was superior to that of today. My formal education, though curtailed since it stopped at the age of fourteen, was good in the three Rs; and while I may not understand the finer points of English, I know how to write grammatically and how to spell, and I can do simple arithmetic in my head.

I regard work as a vocation. I have never been sorry for myself nor do I bear grudges. It pays to look forward and only to take a backward glance to avoid making the same mistake twice. Never be envious of others.

I have often felt that I have been protected in some way in my life – a peculiar and indefinable feeling. I was on two occasions offered seemingly better positions and at the last minute turned them down, to learn later that they proved a disappointment to the incumbents. While possibly co-incident, there have been other happenings which reacted in my favour and for which I have no rational explanation.

While not able to accept the Old Testament picture of an omnipotent and all-seeing God, I believe in a Divine Force which controls through the individual's brain his destiny and, therefore, his actions and approach to life.

I take holidays as a relaxation and not as something one must have because it's the 'done thing'. Our holidays have, in fact, always been pleasant affairs and because Con, my wife, is a Scot, and neither of us like tropical heat, we have spent most of them in Scotland on the Isle of Arran. Here are hills and mountains with trees at the lower levels and heather and fern and, of course, the sea. There is plenty of bird and animal life with the occasional pair of golden eagles, buzzards, hawks and deer. We have many good friends on the island and it is a pleasure to go into the small island shops, be greeted cheerfully and have conversations about this and that; a contrast to the indifference often encountered in London and big city shops nowadays.

Being self-educated and having spent some time on the shop floor, ill-paid as we were then, what I see today is nothing short of appalling – many scrabbling for what they can get, but not willing to give hard and disciplined effort.

When the British economy began to fade after World War Two, it led to a fall in the value of Sterling. Private investment dropped and, because of the post-war Labour Government's nationalization plans, this affected many industries. Nationalization was, and is, along with penal taxation to pay for successive Labour Governments' takeovers and incursions into public ownership, the real reason for our failure to create national wealth. A country based upon Socialism and the Welfare State can hardly become wealthy. I am referring to national wealth for such purposes as housing, roads and up-to-date plant and equipment for factories. These cannot be obtained with sloppily run and over-staffed nationalized (or other) industries.

British Leyland could be regarded as a joke were it not a national disaster and a disgrace. It was a failure before Lord Ryder 're-organized' it. Now, one hopes, by the time this book is published, it may be truly re-organized in the 'vertical' sense, where the different marks or makes of car are run by separately organized companies and under their original brand names, like the organization of the General Motors Corporation in the U.S. or I.C.I. in the U.K.

Everyone in the Government appears to be 'waiting for Godot', or when North Sea oil begins to flow in a big way. But the oil has proved expensive to get and will not make us work harder if used only to back up pay packets without demanding, as a condition, greatly increased human effort. Though feeding on our own oil for power, we should not forget that North Sea oil *is* very expensive and that Middle East oil is, or could be, relatively cheap even after being transported here. So, to stretch our oil treasure for more than thirty years, we might have to consider some of the cheaper Middle East oil, always assuming we have the dollars to pay for it!

We should be depending upon manufacturing those specialist products at which we are good, such as aero engines and, eventually, the motor car, for boosting foreign export sales. Since aviation is the more disciplined in product creation than almost any other industry, we must see to its successful exploitation. It is not good enough to sell, say, a few hundred RB211 engines. A great number, perhaps as many as 4,000, have to be sold to pay the continuing development costs. We will only be able to do this if we can produce an engine markedly superior to that of our competitors and get it installed in a number of commercial aircraft.

In about ten years' time, it could be acknowledged that Rolls-Royce were right in forcing an entry into the American market. But this will only come with future numerical success.

We should not build ordinary run-of-the-mill cargo ships which other countries, like the emerging nations with lower wage levels (e.g. Korea) can be taught to do profitably. Britain must concentrate on specialist vessels like fast naval patrol ships and missile carriers, etc. These demand the highest technologies in engines, hulls and equipment.

We must have more engineers, training them not only in mechanics and physics, but also in plant and works (production) management. There is, thank Heaven, some apparent falling off in those enrolling for social sciences' degrees, but there has been a dearth of trained engineers since the end of World War Two. This is because of the mistaken idea that the advance of the sciences during the war pointed to an apparent need for more scientific training when peace returned. But our real need today is for solid engineering in all the industrial fields, combined with management training. We do not need all the universities there are in this country and some could, to good effect, be turned into advanced schools or colleges of engineering.

Finally, I do not believe in bigness as such, size for the sake of size. We have seen too many amalgamations and takeovers in the last decade in the so-called interests of operating economy and marketing efficiency with the conglomerates resulting from the merging of widely different kinds of companies. Many of these amalgamations have not given the return on capital they were expected to earn, although sometimes the capital has not been sufficient.

While there are obviously certain businesses that require more capital to create greater output at more economical levels, there are conversely many that have become as large and bureaucratic as Governments and that cannot claim better results in specific profitability, i.e. higher output or profits from lower production costs. Much as the Left-wing political idealists decry profitability, this is, after all, a measure of efficiency. These people cannot apparently forget Keir Hardie and his cloth cap. But the difference in his time was that wages were too low and strikes were the only means to redress outrageously poor living conditions. Today, however, the strike weapon is too often used irresponsibly by those who could not care less for the consequences. This is apparent from the fact that few of the strikes start officially and most are lightning affairs which then drift on, with deepening resentment and resistance on both sides. When one hears that a miners' leader, after the October 1974 General Election, is reported to have said that had the Conservatives won, they, the miners, would have called a strike, it is hard to understand such a mentality. If true, it is just further proof of the anarchy being engendered by power lust and a disregard for social order.

Some leaders of the largest and more influential unions are so concerned with politics and their political image that they seem indifferent to the real interests of the mass of their own members. But the way we are going today, with excess personal and corporate taxation offering no incentive to the free investment of money, over-population and the likelihood of up to two million unemployed, there must be a move away from the soft and unrealistic policies adopted since the war. Little or no discipline in the home or in school, softness in dealing with criminals and hooligans, no firmness or organizational ability in handling the Health Service, disorganization of the Steel Industry, all contribute to our low economic state; and one can only plead that we will wake up one day and fight to become a proud nation again.

Having stated why, in my opinion, the U.K. has for some years entered a 'slough of despond' it is perhaps opportune to remind the reader of but a few British achievements. There was the invention of the first steam turbine, the world's earliest television service and the first television sets, radar, the first practical gas turbine engines, the hovercraft, the first supersonic airliner, a successful series of Formula 1 racing cars, the engines of which have kept the U.K. in the forefront of motor racing for about two decades. All this, and much more in activities outside engineering and electronics, must confirm the lively brainpower and inventiveness of our countrymen. That we can collect ourselves, recover and become great again is certainly possible. But over the past two centuries much of this inventiveness has passed from our little island to Europe, the U.S.A., and Japan, in fact around the world. That inventiveness, still a part of the British way of life, is in my view unchallengeable but its fruits will elude us unless we take a pride in what we can do; and unless we do as much for our country as we are prepared to do for ourselves.

It can be done – but will it be done?

Author and Dennis Head of Rolls-Royce standing before the Supermarine S6B racing seaplane with the Rolls-Royce 'R' engine between them. The occasion was the presentation of the Schneider Trophy to the Science Museum on 15 March, 1977. By permission of the Science Museum, London.

John Cobb (right) and Reid Railton at Salt Lake, Utah, U.S.A. after their car had reached more than 400 mph on the Author's special fuel mix. (16 September, 1947).

With the late President Nkrumah of Ghana after presenting him with a silver model of his Presidential HS125 which had just been delivered to Accra.

Prince of Wales with the Italian Team at Calshot, 1929. Looking at photograph left to right: Lieut. Cadringher; Lieut. Monti; Capt. Canateri; Prince of Wales; Lieut. Col. Bernsconi; Warrant Officer Dal Martin and Sergeant-Major Agello.

A visit to Mr Honda at his plant in Japan, January 1964.

Appendix I
French Higher Education

The basis of French education is the baccalauréat, an examination taken when school children are about seventeen. There are various forms of baccalauréat, denoted by the letters A, B, C, and so on, but for those who want to go on to an advanced course in scientific studies, baccalauréat C is generally a 'must'.

Having passed the examination, the student has a choice. He can go to university, as in Britain or the U.S., where after four years he is generally awarded a bachelor's degree (in France, 'Maîtrise'). The baccalauréat is sufficient qualification for entry into a university – there are no special entrance examinations – but a number of undergraduates tend to leave in the first year, and sometimes even in the second. After his first degree course, the student can go on to take a master's degree ('Doctorat'), which usually takes three years. Graduates nowadays often have difficulty in finding jobs in industry; many become teachers in schools or universities, or research technicians in State laboratories (Centre National des Recherches Scientifiques, Commission a l'Energie Atomique, etc.).

The other alternative is to try for entry to a 'grande école'. Some thirty years ago the number of 'grandes écoles' was relatively small, but since the war they have expanded and degree courses in engineering are available at many of them.

For entry into a 'grande école' the student has to pass an examination which will entail two or three years' preparatory work. Some of these entrance exams are extremely difficult, particularly those for the Polytechnique (where each year 300 students are selected from 2,000 well-qualified applicants), the École Centrale, the École des Mines, etc. Others are easier. So a boy or girl can choose to try for a 'grande école' where the level of work is suited to his or her capabilities – in fact, most students will try for two or three, to be sure of obtaining a place at one of them. The course at a 'grande école' normally lasts for three years and ends with the award of a 'diplome'.

At the École Polytechnique, courses last for only two years. The Polytechnique was originally created by Napoleon for the education of high-level 'fonctionnaires' and artillery officers. It now has such a high reputation that its former students have no difficulty in finding jobs in industry; few now join the Army, which is not so well paid. At the end of his second year, the student chooses, according to his grade, the 'grande école' which will give him his specialised training. The ten best students generally choose the École des Mines, and others Ponts et Chaussées, Telecommunications, Genie Maritime, Aeronautique, etc. After this one-year specialization, a graduate is known as an 'Ingenieur du *corps* des Mines', as distinct from a graduate who has spent three years at the École Superieure des Mines de Paris, for instance, who is called an 'Ingenieur des Mines'. This distinction is an all-important one in France. Many of the most prestigious jobs are held by graduates of the Polytechnique, or the École Nationale d'Administration (for finance and economics). There is considerable rivalry between the two institutions.

In the petroleum industry, all the top jobs at C.F.P., C.F.R., Antar, Elf, etc. are held by École Polytechnique graduates, particularly those engineers who have spent one year at the École des Mines. In a French company such as the Compagnie Francaise de Raffinage, even if the Managing Director is 'du corps des Mines', the top managers are *all* from the École Polytechnique and have generally specialised at Ponts et Chaussées or Génie Maritime. An engineer who has trained and taken his degree at, for example, the École Centrale, without the two-year course at the Polytechnique, will find it difficult to reach the top; he will always be more or less limited.

Appendix II
Fuel

While I specialized in the application of fuel to engines in the between-wars period, I do not claim to have any particular knowledge today; and while I have naturally mentioned fuel in this book, I thought a little more history would be valuable in view of the extreme importance of fuel to the piston engine in the war (see reference [a]).[1]

When we heard of the American advances in the manufacture in quantity of 100 octane fuel by the hydrogenation and then the alkylation processes it was clear to many of us that this was obviously the fuel for the Royal Air Force. Already, in 1938, 90 per cent of the U.S. Government purchases were for 100 octane fuel, whereas British fuel for the R.A.F. was 87 octane grade, under Air Ministry specification DTD-230.

I gave a paper in London in January, 1937, and pleaded for the development of British military engines to take 100 octane fuel. It was criticised by a high oil company official on the grounds that we might not get 100 octane in war time and should therefore opt for 87 octane. Admittedly, 100 octane was only obtainable in the U.S. at the time; and if America was not in the war, her Neutrality Act could prevent supplies; though Anglo-Iranian, among other oil companies, was working on the alkylation process.

Fortunately, the critics were unheaded and in 1937 the Director of Technical Development of the British Air Ministry agreed to develop engines to run on 100 octane fuel. So by 1940 the Merlin's power was increased by this fuel from a combat rating of 1,000 (plus) bhp to over 1,300 bhp, and 100 octane became available to Fighter Command ready for the Battle of Britain through Roosevelt's 'cash and carry' compromise. It was the alkylation process, introduced in 1938, that met the rapidly rising needs for 100 octane fuel. While the paper mentioned in footnote (a) gives a lot more detail for those directly interested, it was the supply of 100 octane fuel for bench and flight tests which was very critical and important, and to one man can be credited the foresight and the energy to get a tanker load of this fuel delivered to the U.K. This man, Dr Bill Sweeney of Esso, blended the constituents of the hydrogenated 100 octane fuel to meet the British requirements for good 'rich mixture response'. Sweeney's fuel blend was first tested at the Army Air Corps (later U.S.A.F.) experimental and test base at Wright Field in single cylinder units and later in our own air-cooled (Pegasus and Mercury) units, which were more sensitive than the water-cooled Merlin. 'Sweeney's Blend' was a great success and in June 1939, only three months before the start of World War Two, a company tanker, the *Beaconhill*, set sail across the Atlantic for the U.K. with a full cargo of 100 octane fuel. We began stockpiling this fuel, though the decision to use 100 octane for Fighter Command was not made until March 1940 and that for its use by Bomber Command came in 1941.

I had early manifestation of 'rich mixture response', quite accidentally I might say, when preparing the benzole-alcohol fuel for the 1931 Schneider engines. This is acknowledged in the Schlaifer and Heron Book, *The Development of Aircraft Engines and Fuels*.

There was, on occasions, a special fuel component blended into 100 octane for Merlin engines in the Mosquito, to accelerate away from enemy fighters over Berlin or to catch the VI flying bomb. This was mono-methyl aniline, of which about 8 per cent (plus) was added.

The aviation gas turbine engine is more catholic than the piston engine in the fuel it can use, though kerosine (paraffin) is its normal fuel (dewaxed to lower its freezing point to about minus 40°C). It has, however, been found that kerosene, having an aromatic content of 20 per cent and above, can cause an increase of 8°C (14°F) in combustion liner temperature compared with that of standard Jet A fuel of 15 per cent aromatics (see reference [b]).[2] There is, normally, no detonation problem in a continuous combustion process engine like the gas turbine.

For the benefit of those readers interested in a simple explanation of the subject, Ethyl Fluid consisted of tetraethyl lead $Pb(C_2H_5)_4$ and a halogen bearer such as ethylene dibromide or ethylene dichloride. The latter was added to the lead in amounts of 1 or 1.5 theories (one theory being the theoretical chemical amount of halogen bearer fully to convert the lead oxide formed during combustion to more volatile compounds (lead bromide or lead chloride) for the better evacuation from the engine cylinders, to reduce deposits.

Ethylene dibromide was favoured for the Ethyl Fluid used in aviation fuel since it was more effective (volatile), but more expensive than ethylene dichloride – which latter was used for many years in

MOTOR MIX with a half theory of ethylene dibromide for automotive gasoline – $1\frac{1}{2}$T total. I-T AVIATION FLUID contained tetraethyl lead with one theory of ethylene dibromide. In recent years, tetramethyl lead (T.M.L.), a more volatile form of lead than T.E.L., has been used with the latter, to improve lead distribution and anti-knock effect throughout the fuel particularly in the lighter fuel ends.

Ethylene dibromide is manufactured by processing vast quantities of sea water. Bromine is present in sea water in the proportion of about seven thousandths of one per cent, but one cubic mile of sea water contains about 600,000,000 lb of bromine.

Lancaster Mark I.
   Used Merlin XX, 22 or 24 engine

Lancaster Mark III & X
   Used 28 or 38 engine

Packard Merlin Engine
   28 & 38 (equivalents XX & 22)
   All 3000 MAX RPM
   @ 9 PSI Boost

[1] [a] *Milestones in Avaition Fuels* by W. G. Dukek, D. P. Winans and A. R. Ogston. Paper given at A.I.A.A. Designers and Operators meeting, July, 1969, Los Angeles.
[2] [b] *Looking for Fuels in the Future* by William Dukek and Dr. John Langwell. Petroluem Review, April, 1978.

Appendix III
Engine Production

A few brief remarks on the production of aero engines will suffice here, since a good deal of valuable information is given by Devons in his book *Planning in Practice*.[1]

Rolls-Royce Merlin
There were three broad groups of Merlin engines: the early Marks from 1 to 21, produced in the U.K., were single-stage supercharged engines with the single-piece cylinder block, followed by a group of single-stage engines having a two-piece cylinder block. The single piece block suffered glycol leaks and was a danger to fighter pilots. Here the liner was held in compression by throughbolts while the two-piece block had the flange at the top of the cylinder liner nipped between the cylinder head and the main block and was a much better job, giving a more coolant proof joint. The two-piece block engines, Mks 22, 23, 24 and 25, were British productions and the Merlin 38 and 224 U.S. (Packard) productions for bomber aircraft. Marks 30, 32, 45, 47, 50, 55 and 56 (British production) were for fighter aircraft. Then came the development of the two-stage supercharged Merlin (61, 63, 64, 65, 66, 67, 70, 71, 72, 73, 76 and 77) produced in the U.K. for fighters and the Mark 68 (U.S. production) and 85 (U.K. production) for bombers.

The British Ford Company helped greatly in engine production, when Sir Wilfrid Freeman asked Rowland Smith (now Sir), then Managing Director of Ford U.K., if he would produce Merlins to supplement Rolls' production. They agreed with Hives to set up a factory at Government cost to produce about 400 Merlins a month. At this time, in Beaverbrook's early days as Minister of M.A.P., there was another top Ford man in the Ministry seeing to production facilities. He was Patrick Hennessey (now Sir).

In round figures there were 2,600 Merlins produced in the first 12 months of the Second World War and by 1943 the combined Derby, Crewe and Glasgow factories were turning out Merlins at the rate of 18,000 a year. The 100,000th Merlin engine (including Ford, U.K. and Packard, U.S.) was produced in March, 1944; and over 165,000 Merlins had been produced by the end of the European War.

There was, of course, the 'Griffon 65' engine, the production history of which is well-described by Devons. It was intended for the Spitfire and Spiteful, the airframes being somewhat behind the production of the Griffon, which had been ordered ahead in the thought that it would lag behind the airframe. In the event the reverse happened.

Bristol Hercules, Centaurus and other engines
(a) Hercules XVI and 100
These two engine Marks were fitted to the Halifax and Sterling bomber aircraft, and as the engine programme stood at the time there were over 3,000 surplus Hercules at the end of 1945.

A doubtful element in the Hercules programme had been the future of the Sterling, which seemed to drift on in production at Short Harland. There were four factories in the Bristol Hercules Group – Bristol (Parent), Bristol (Accrington), Bristol (Hawthorne) and the Shadow operations. The underground factory at Corsham (Hawthorne), originally pressed into being due to the efforts of Roy Fedden, had a relatively poor production record on the Centaurus engine and it seemed that Bristol's Accrington factory would have to take on its production along with its Hercules output. During the period 1939 to 1945 the Bristol aero engine factories and the various shadow factories produced the following radial engines:–

Mercury	32,000
Pegasus	32,000
Hercules	65,000
Taurus	3,400
Perseus	8,000
Centaurus	2,800

This represents a grand total in excess of 143,000 engines.

Napier Sabre
In its early production the Sabre was in great shortage, largely due to its high maintenance rate which resulted in a large number of engines undergoing repair or awaiting repair. This engine was wanted only for the Hawker Typhoon and Tempest fighters, but by about 1944 its reliability had been greatly improved to the point that, given the normal allowance for spares, there was in the 1945 engine programme a surplus of some 1,500 engines. The Sabre was built at the parent factory at Acton (London) and the Walton factory (Liverpool).

[1] *Planning in Practise* by E. Devons, Cambridge University Press, 1950.

Index